建筑教学实践

——北京751地块三维城市结构设计

[芬] 卡斯·欧斯特豪斯　托马斯·亚斯凯维茨　编
冯瀚　张愚　译

中国建筑工业出版社

图书在版编目（CIP）数据

建筑教学实践——北京751地块三维城市结构设计/［芬］欧斯特豪斯，亚斯凯维茨编；冯瀚，张愚译. —北京：中国建筑工业出版社，2008
ISBN 978-7-112-10281-5

Ⅰ. 建… Ⅱ.①欧… ②亚…③冯…④张… Ⅲ.城市规划—建筑设计—北京市—高等学校—教学参考资料 Ⅳ.TU984.21

中国版本图书馆CIP数据核字（2008）第122340号

责任编辑：唐　旭　姚丹宁
责任设计：董建平
责任校对：兰曼利　孟　楠

建筑教学实践——北京751地块三维城市结构设计
［芬］卡斯·欧斯特豪斯　托马斯·亚斯凯维茨　编
冯瀚　张愚　译
*
中国建筑工业出版社出版、发行(北京西郊百万庄)
各地新华书店、建筑书店经销
北京图文天地制版印刷有限公司制版
北京画中画印刷有限公司印刷
*
开本：889×1194毫米　1/20　印张：8　字数：240千字
2009年1月第一版　2009年1月第一次印刷
印数：1—2500册　定价：58.00元
ISBN 978-7-112-10281-5
(17084)

Foreword
前言

798艺术区位于北京东北部。这里以前是个电子器件工厂，该区西部最近已发展成一个充斥着画廊、展示、餐厅和创意办公的活跃区域。它是自下而上形成的。与世界上其他后工业地段相似，一个个艺术家发现了这些令人激动的废弃空间，并将其改造出新的使用功能。这最终使798变成了城市最时尚的区域之一，充满了生机和活力。目前，798艺术区正在启动进一步开发，吸引更多的用户和功能。该区西部已成功改造，东部仍然在等待新生。

798艺术区需要进一步扩大。它需要为新的功能和使用者提供新的空间。它要成为北京城市景观的主要焦点之一。然而，如果认为一个自上而下的设计，就能满足正在成长的798社区的所有需要，那就错了。该区域的扩建必须是弹性、多样和不受约束的，而且必须成为这个惊人地段的地标。

作为对此需要的回应，Kas Oosterhuis教授和Hyperbody共同提出了一个想法，来为798区域的东部进行分布式的设计——751工厂。这个概念是要建立一个大尺度、多功能、三维的城市结构，作为自下而上、分布式设计过程的结果。通过把该区域划分为更小的、相互依赖的项目，分配给自治、但同时又密切合作的设计者，从而获得自下而上开发的本质。一方面，他们都应拥有其所需的设计自由，另一方面，其设计过程作为群体行为应当是个游戏，遵循相同而明确的规则设置。

The 798 art district is located in the northeast part of Beijing. Formerly an electronics factory, its western part has recently developed into a vibrant district, full of galleries, exhibition spaces, cafes and creative offices. It has emerged from a bottom-up process. Like in many other post-industrial areas all over the world, individual artists found exciting, unused spaces and adapted them for new functions. This eventually turned the 798 into one of the most hip zones of the city. Already full with life and activities, the 798 art district is bound to develop further and to attract many more new users and functions. The western part of the area has already been successfully transformed. The eastern part still awaits to be re-filled with life.

The 798 district needs to grow further. It needs new spaces for new functions and for new people. It needs to become one of the main focal points in the cityscape of Beijing. Nevertheless, it would be a mistake to assume that one, top-down design can provide an answer to fulfil all needs of the growing 798 community. The expansion of this area has to be flexible, diverse and unconstrained and it has to become a landmark icon for this amazing site.

As an answer to this demand professor Kas Oosterhuis together with the Hyperbody came up with an idea of making a distributed design for the eastern part of the 798 district - the 751 factory. The concept was to create a large scale, multifunctional, three dimensional urban structure, which would be conceived as a result of a bottom-up and distributed design process. The bottom-up nature of the development can be achieved by dividing the area into smaller, interdependent projects, assigned to autonomous but at the same time closely cooperating designers. On one hand all of them should have all the design freedom they may need, on the other hand their design process as agroup should be a game in which they all follow the same, clear set of rules.

Table of contents
目录

Hive Mind
蜂群思维

prof. ir. Kas Oosterhuis *

751项目规划设计是以一个明确概念和有限的几条强制规则为基础的多用户设计游戏。多用户设计的目标在于有效增强小组设计者之间的合作。这种目标一旦达成，那么设计过程将会自下而上地控制，促成总平面的多样统一。

蜂群思维

首要任务是要找到一个概念，以促成多样统一的发展。一个人或持单一思想的小组设计整个城市是极端有问题的，城市多样统一的概念则可潜在地避免这种情况。单一思想导致环境精神品质低劣的例子在过去太多了。人们对于单一思想控制的规划很反感："过于雷同"、单调、视觉贫乏。身处此种城市景观中的居民在精神上缺乏足够的社会和形象刺激。而个体思维的松散组织也不会让城市正常运转。这种城市无异于独立设计、彼此毫无联系的孤岛群。避免这种单一思想的可能方法与它的作用一样简单。总平面应当由相互联系的思维来设计和实施。设计者的思维必须相互联系。相互联系意味着个人努力必须基于他们所共享的东西，但并非决定其使用功能。可以想象，各个用户通过互联网相连，从而形成一个社区、一个群集。在这样一个互联网社区中相互联系着的人们，从属于群集成员，而该技术并不决定其使用功能。互联网技术将这些用户一起带入一个蜂群，这充分描绘了这种可避免单一思维的坚决状态：蜂群思维。

The 751 masterplan is based on an explicit concept and a limited number of strong rules for a multiplayer design game. The aim of multiplayer design is to improve the effectivity of collaborations between a group of designers. Once the effectivity has been improved the design process will be bottom-up controlled as to facilitate the multiple identity of the master plan.

Hive mind

The first task is to find a concept which facilitates the evolution of a multiple identity. The concept of multiple identity for a city potentially solves the extremely problematic condition of one single person or one single-minded group designing a complete city. There are too many examples from the past that single-mindedness causes emotionally poor environments. The users will experience a single-minded masterplan as boring, "more of the same", monotonous, visually underdeveloped. The inhabitants in such a single-minded cityscape are emotionally deprived from sufficient social and formal stimuli. A loose group of individual minds would not be able to make the city work. That city would be no better then an archipelago without any relations between the individually designed islands. The possible solution

总体框架

在设计小组开始设计像城市那样错综复杂的东西之前，他们必须展开这种蜂群思维。理论上说，这能够产生于自下而上的过程，确切来说需要很长时间。我们不能强求进化本身的速度，在时间上没有可能的捷径。我们不得不首先建立起总体框架。我们必须设定这个设计游戏的规则。所有规则的设定形成了初始状态，蜂群就居于其中。初始状态可以与互联网相比较。它是所有用户都必须遵循并用来建立彼此联系的规则，他们将用这些规则来游戏。游戏者若违反规则，则会与蜂群断开联系，正如互联网用户如果不使用正确的软件或不遵循网络供应商的指导，就将与网络断开连接一样。接入网络就意味着要接受连接系统的内在规则。要记住，这些都不是关于使用功能的，而只是设备技术。是否使用这些技术在此语境下是不相关的。与此类似，谁被授权参与该总体规划设计的游戏也不在751项目，或者其他任何此种项目范围之内。

总体框架的规则

总体框架意在用最少规则产生尽可能强的效用，这是效用的本质。所谓效用是一种必要状态，能够进行透明的多用户游戏设计过程，并产生尽可能有意义的复杂性。复杂并非繁杂，繁杂是无休止违反规则的结果，而复杂来自于简单的规则。总体框架形成于一个盘旋于751基地上空的工作空间，它是一个具有假定形状和尺寸的巨型球体，该球体形成了总体框架的边界状态。边界状态是自上而下决定的，设计者们要将其接受为“既成事实”。边界状态决定了后续设计过程在几何位置上从何处开始与结束。该球体细分为23个三维地块，精致地相互交织，形成了一个三维智力游戏。每个设计者只在其中一个小地块上展开设计。各设计者分别接受作为既成事实的各三维地块，正如在城市肌理中他们也将接受一个二维的城市街区一样。但是现在他们有着更多的相邻地块，不像在城市平面上那样只有两个或四个相邻地块，而是可能有高达10个地块与其相连。设计者的给定任务只是与其直接相连的地块沟通。设计者被明确告知不用建立整个球体的概念。规则要求他们只需与相邻地块沟通，就像群集中的鸟儿一样。这就是他们将如何建立蜂群思维的过程。蜂群思维没有针对整体形状和尺度的线索。任何单独的互联网用户都不会知道互联网究竟是如何延伸出去的。用户只对其同类个体感兴趣，并通过网络技术与他们交换数据。群集中的鸟儿在同类个体之间相互沟通，而在751总体框架中，设计者也是在同类个体之间相互沟通。当数据流出一个地块时，751项目总体规划的设计者，即TU Delft Hyperbody的硕士生们，必须通知其相邻地块。有数据进入自己的地块时，他们也将被相邻地块告知，就是这些。数据被限制为结构力、水流和电流、人流和车流。此外，各种需求功能的数量也给这些设计学生规定了如居住、办公、商业和停车的面积。最后一条规则是，其地块最多只能有25%的地方用于建造房屋。这就意味着这个体量巨大的三维球体在城市结构上将会是多孔渗透的。

to avoid single-mindedness is just as simple as it is effective. The masterplan should be designed and executed by connected minds instead. The minds of the designers must be connected. Being connected means that the individual efforts must be based on something they share, but without determining the content. Think of individual users, connected via the Internet forming a community, a swarm. The subjects which are communicated in such an Internet community are subject to the members of the swarm, the technology does not determine the content. The technology of the Internet brings these users together in a hive, describing adequately the mandatory condition for avoiding single-mindedness: the hive mind.

The Master Frame

Before the groups of designers may start on the effort of designing something complex and intricate as a city they must develop that hive mind. Theoretically this could arise from a bottom-up process, but this would literally take ages. We can not evolve evolution faster then evolution itself. There is no by-pass possible through time. We have to construct the Master Frame first. We must set the rules of the design game. The complete set of rules forms the Initial Condition, where the hive lives in. The Initial Condition can be compared with the Internet. It is a set of rules which all users obey and use to build their connections. In much the same sense the Master Frame is defined through a set of rules which the designers of the Master Plan will obey, they will play by these rules. Violating the rules will simply disconnect the players from the hive, just like the Internet users will be disconnected if [s]he does not use the proper software or obeys the instructions as set by their providers to get access to the Internet. Getting access means accepting the rules which underlay the connectivity system. And remember this is not about content, it is only facilitary technology. The question whether or not to pay is in this context not relevant. Similar to the question who will be authorised to access the Master Planning design game is outside the scope of the 751 project. This typically concerns a pre-selection process preceding the 751 project, or any other project in this respect.

Rules of the Master Frame

The Master Frame has been defined as to give the minimum amount of rules with the strongest possible effect. This is the essence of effectivity. Being effective is a necessary condition to be able to embark on a transparent multiplayer design process resulting in the highest possible meaningful complexity. Complexity rather then complicatedness. While complicatedness is the result of an endless series of exceptions, complexity arises from simple rules. The Master Frame is formed within the boundaries of a working space with the hypothetical shape and dimensions of a giant sphere, hovering above the 751 site. The sphere forms the boundary condition of the Master Frame. The boundary condition is a top-down decided fact which the designers will accept as a "fait accompli". The boundary condition decides where the subsequent design process begins and ends geometrically. The sphere is subdivided in 24 3d plots, delicately interwoven as to form a 3d puzzle of plots. Each designer will develop the design of only one such a plot. The 3d plots are a given fact and accepted by the individual designers, just in the same way as they would accept a 2d plot city block in the city fabric. But now they have more neighbours. Not only two or four as in the 3d city plan, but up to 10 neighbours may connect to their plots. The task the designers are given is to communicate with their immediate neighbours only. The designers are explicitly asked not to have a concept on the whole sphere. The rule is that they will communicate with their immediate neighbours only, just like the birds in a swarm. This is how they will develop the hive mind. The hive mind has no clue to the shape or dimensions of the whole. No single Internet user knows how extended the Internet really is. The user is only interested in their peers, with whom they exchange data through the technology of the net. Birds in a swarm communicate peer to peer, and in the 751 Master Frame the designers communicate peer to peer. The designers of the Master Plan, in the 751 project the Master students of Hyperbody at the TU Delft, must inform each other about the data flowing out of their plots. They will be informed by their neighbours of the data coming into their plot, and that's all. The data are confined in terms of structural forces, water and electricity flow, and people / car movements. Furthermore

自下而上的总体规划

在接受了总体框架规则后，这些规划师们开始了他们的设计工作，设计游戏启动了。在此过程中，学生们可以与相邻地块进行土地交易，只要用地总建筑面积和整个球体的总建筑面积得到满足。球体总建筑面积是一个需要满足的总体参数，它只决定于总体框架的变化，而不是任何一个规划师。使用功能不能忽视，总居住量和其他使用功能必须保持不变。23个学生将发展各自的设计主题，这些主题可能非常不同，但又相互联系。这一观念完全显示了总体框架概念的力量。这些联系在一些明确规则的基础上一旦建立，逐渐展开的复杂性过程就开始了。视觉多样性和不同景致逐步展现，并仍保持着相互联系。如果没有这些规则的设置，而是接受各参与者不同的设计态度，那么将导致巨大的冲突。各地块设计者的唯一职责就是根据其自己的设计标准来开发用地，他们始终保持与紧邻地块联系，并交流数据的变化。数据交换是通过一个内在的数据库结构来组织的，它记录了所有数据的变化。设计者们是在此过程中体验工作，而不仅仅是谈论此过程。

在进化中工作

于正在进行的过程中工作，这是由不断变化的动态数据库支配的，就像在进化中工作。这一观念对基于规则的设计来说非常重要。规则一直在执行，永不停息，就像进化本身也是一个不可停止的过程。不论你是否喜欢，你的生命都在进化中展开。就像在751设计游戏中的设计者们一样，人们对于进化可进行外在观察。我们深陷在内部，但这不是问题。相反，在进化中生活，意味着接受和执行生命游戏的规则，这让我们兴奋，因为这表明我们在世间的存在确实很重要。生命只不过是游戏者之间的述行交互。生命包括产品的连续设计和生产，通过产品生命的进化，生命得到扩展。我们这些设计师为产品生命的进化提供帮助。我们移动物质，我们移动数据。人是物质形式和其他形式的数据的传输者。人是数据的递送者，包装、运动、再拆包。不要出错，这是一个人可能期望的生命最深刻的意义。进化之所以成为进化，因为它在实时进行，每个最小的时间单位都完成了数十亿次的离散步骤。生命，就像751中的设计过程，是一个运转过程，一个复杂的运转着的细胞自动机，组成细胞自动机的每个单元基本上都有一个实时执行的简单规则。751项目的学生设计者就是进化的操作者，他们操作着生命游戏的规则，进而操作了751设计游戏。通过玩这个设计游戏，他们实时表现为群集的可靠成员，导演了总平面的进化，他们共同形成了751的蜂群思维。

* Kas Oosterhuis，交互建筑学教授，TU Delft建筑系Hyperbody负责人，并开设鹿特丹ONL [Oosterhuis_Lénárd]设计事务所。

the student-designers are given a numeric program of demands with number of m2 for residential, offices, commercial areas and parking. In addition to that the last rule is that a maximum of 25% of their plot may be built-up. Meaning that the large volume of the 3d sphere will be quite porous in its urban structure.

Bottom-up Master Plan

Having accepted the rules of the Master Frame the Master Planners start their design work. The design game unfolds. In this process the students are allowed to trade m2 with their neighbours, as long as their plot GFA and the sphere GFA is respected. The sphere GFA is a global parameter to be respected, and only subject to change by the Master Framer, not by any of the Master Planners. They will never be allowed to skip content, the total mass of residential and other programmatic content must remain constant. The 24 students will develop their own design themes, they may be very different. Very different but connected. This notion describes properly the power of the concept of the Master Frame. Once the connections are established on the basis of some explicit rules, the process of unfolding complexity starts. Visual diversity, different views unfold yet stay connected. If we would not have these rules set and accepted by the players the difference in the design attitudes would lead to enormous conflicts. Not here since this is something they simply not talk about. It is the sole responsibility of the plot designer to develop their plot according to their own design standards, as long as they stay connected and communicate the changes in data with their immediate neighbours, all the time, in real time. The exchange of data is organized in an underlying database structure which accounts all changes in the data. The designers experience working inside a process, rather than talking about a process.

Working inside evolution

Working inside the ongoing process, which is administrated by the dynamic database of ever changing data, feels like working inside evolution. This notion is crucial for rule-based design. Rules are always running, they never stop. Like evolution itself is a process which is unstoppable. Whether you like it or not, your life unfolds inside evolution. Like the designers in the 751 design game, people have exterior view on evolution. We are trapped inside, but this is not a problem. On the contrary living inside evolution, which means accepting and executing the rules of the game of life, provides us with the excitement that our presence in the Universe does matter. Life is nothing more then the performative interaction between its players. Life includes the continuous design and production of products. Life is augmented with the evolution of product life. We designers assist in the evolution of product life. We move matter, we move data. I would not hesitate to see this as the meaning of the evolution of people in the Universe and hence in their bi-directional relation to products. People are transporters of matter and other forms of data. People are carriers of data, packing, moving and unpacking data. Don't get it wrong, this is the deepest meaning of life one could possible hope for. Evolution is evolution since it evolves in real time, taking billions of discrete steps per the smallest time unit. Life, as is the design process inside 751, is a running process, a complex set of running cellular automata, while each of the constituting cellular automata basically is a simple rule, executed in real time. The student-designers of the 751 project are such operators on evolution, operating on the rules of the game of life, and hence on the design game of 751. Playing the design game they evolve the Master Plan, behaving in real time as reliable members of the swarm. Together they form the hive mind of 751.

* Kas Oosterhuis, professor Interactive Architecture, is the director of Hyperbody at Faculty of Architecture, TU Delft and runs the design office ONL [Oosterhuis_Lénárd] in Rotterdam.

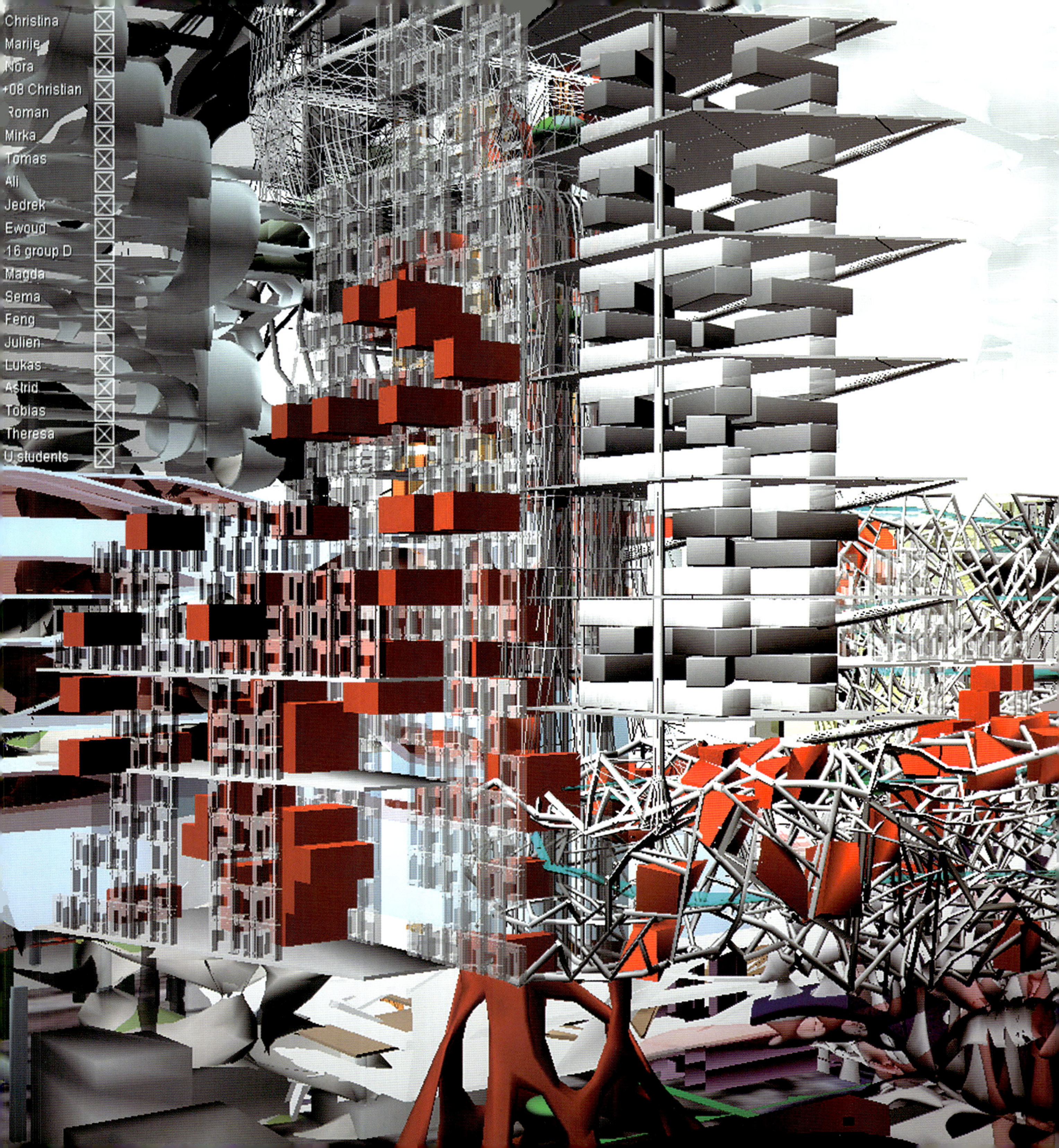
Christina
Marije
Nora
+08 Christian
Roman
Mirka
Tomas
Ali
Jedrek
Ewoud
16 group D
Magda
Sema
Feng
Julien
Lukas
Astrid
Tobias
Theresa
U students

Multiplayer Design
多人设计

Tomasz Jaskiewicz

设计建筑是一种游戏。在这项游戏中我们遵循规则。那些规则当中许多已经为我们预先确定了，余下的则由我们自己开始这项设计游戏时来确定。我们通过达到我们的目标来赢得这场游戏，我们也和其他人一起参与这场游戏。如果我们的目标相冲突，我们则相互竞争。如果我们的目标相一致，我们合作。

由代尔伏特理工大学的Kas Oosterhuis 教授和Hyper-body* 工作室组织的751合作设计课程的目标是设计一个大胆的、丰富的和多样化的三维城市结构。课程始于2006年9月，包含23位来自全球的学生。在这开始以前，我们已经决定设计的基地是一个包含800万m^{3}容积的三维虚拟球体。

整个学期的目标是形成一个包含真实城市环境复杂性的设计模型，这一模型能够随外界参数进化为不同状态。为了达到这一点，球体被分为23个内部的区域。每一区域分给一位参与的学生。整个虚拟建造选址于北京751工厂的中心，部分位于地下，占地面积是不大的20000m^2，高出地面120m。751工厂剩下的面积分为5个部分，由来自南京东南大学的董卫、张倩、杜蓉老师指导的5位学生设计。他们的工作平行于代尔伏特理工大学设计小组的进程。他们的设计是产生于二维的设计观点。整个设计过程也有来自北京清华大学，由徐卫国教授指导的学生组成的小组的参与。这一小组使用先进技术和理论以变量的方式捕捉整个项目的地方特色。

Designing architecture is a game. In this game we follow rules. Many of those rules are predefined for us, the rest we set ourselves upon starting the game of designing. We can win this game by reaching our goals and we play the game always together with other people. If our goals are conflicting, we compete. If our goals are complimentary, we collaborate.

The goal of the 751 multiplayer design studio organized by prof. Kas Oosterhuis and Hyperbody* at TU Delft was to design a daring, rich and diverse three dimensional urban structure. The studio started in September 2006 with 23 students coming from all around the globe. Before it had begun, we have decided that the group will get as a design site a three dimensional virtual sphere containing 8 million cubic meters of volume.

The objective set for the entire semester was to produce a design model that would have the complexity of a real urban environment and which would potentially be able to evolve into many variants conditioned by external parameters. To achieve this, the sphere has been divided into 23 interlocking zones. Each of those zones has been assigned to one of the participating students. This whole virtual construct has been then placed in the centre of the 751 factory in Beijing. Partly submerged underground, it would stand on a small base of 20 000 m^2 and raise 120m above the ground level. The rest of the 751 area has been divided into five parts and became subject to designs of five students from the collaborating South-East University of Nanjing guided by prof. Dong Wei, Zhang Qian and Du Rong. Their work proceeded in parallel to the progress of the TU Delft design studio. Yet, their designs were conceived from a more conventional, two dimensional design perspective. The whole

代尔伏特理工大学的学生由Kas Oosterhuis教授指导。设计指导Tomasz Jaskiewicz，Gijs Joosen 和 Marthijn Pool与技术指导Gerrie Hobbelman，Bige Tuncer 和 Guus Westgeest一起更近距离地指导学生。整个学期举行了一系列额外的研讨，受邀的讲师包括Jeroen Coenders，Jerome Decock，Lukas Feireiss，Ulrika Karlsson，Chris Speed 和 Patrick Teuffel。学生们通过Bert Bongers的演讲和多媒体演示以及Hans Hubers（Hyperbody的教学负责人）指导的研究与设计课程，拓宽了他们的知识与技能，鼓励每一位学生具有创造力和革新能力。他们被允许发展甚至是最疯狂的概念，只要他们的设计遵循以下的总平面设计原则：

1.功能性的内容是一项100万m^3的混合使用的建设范围。

2.包含在800万m^3的一个虚拟球体内，75%的部分是开放空间，让阳光能够进入这一大的城市体当中。

3.位置需要处于751场地的中心。

4.一个三维的迷团分为许多连锁的部分，学生们每人分得其中一块。

5.每一个这样的三维区域仅同和它相邻的部分交互和对话。

6.这一三维谜团当中的每一部分对应一种功能（居住、办公、商业、文化、教育、休闲）。

7.每一区域管理它们自己的数据输入、数据处理、数据的输出以及通过一个动态的数据库和它们邻近的局域进行数据交流。

8.每一区域具备自我支撑的结构体系并且同相邻部分进行结构数据的交换。

9.球体必须产生与其消费相一致的能量。

设置那些规则的主要逻辑源于群体行为的想法。正如 Kas Oosterhuis教授定义的那样，在一个群当中，每一部分仅同其相邻部分交流。群的整体形状并不是由其中的任一要素制定的，当中没有领导，群的要素也不需要了解整体。群的结果是一个来自内部自发的，双向互动的组成要素以及严密的外部条件的平衡。在751规划中，学生是自发交流的成员，教师则进行自上而下的控制。

根据整体规划划分的三维基地意味着所有设计者之间必然是内部关联的。一些部分不得不由其他部分支撑，另外一些部分需要为那些同外部环境隔离开的部分提供进入的通道。所有部分需要一起工作来处理阳光的射入、人、水、能量和废物循环以及其他问题。

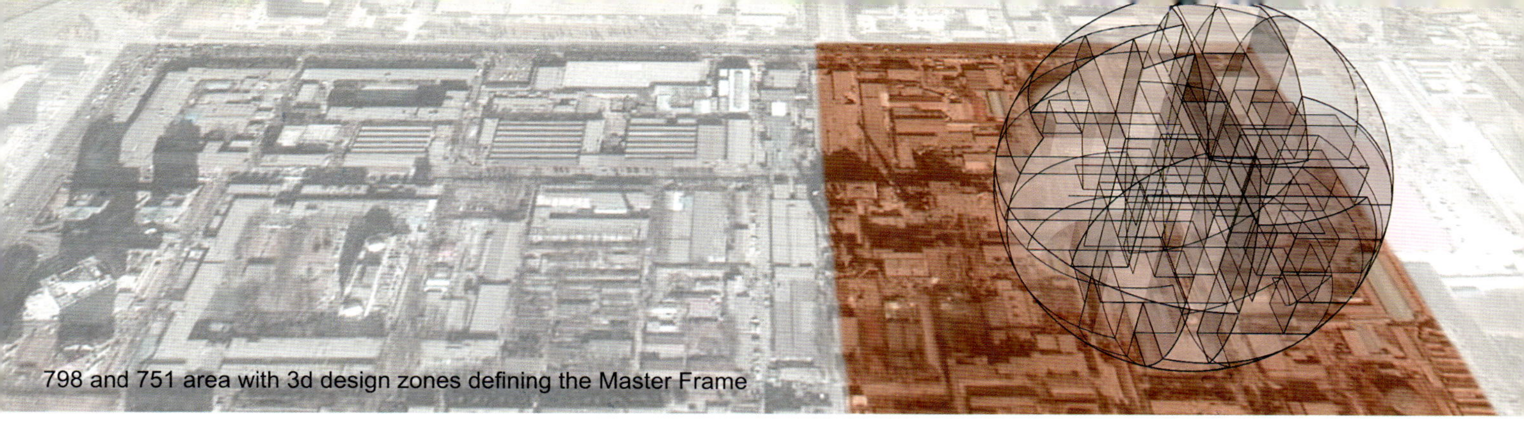
798 and 751 area with 3d design zones defining the Master Frame

design process has been also followed by a group of students led by prof. Xu Weiguo from the Tsinghua University in Beijing. That group used advanced techniques and theories to capture in a parametric way the genius loci of the entire project.

The group of TU Delft students has been supervised by prof. Kas Oosterhuis. The design tutors Tomasz Jaskiewicz, Gijs Joosen and Marthijn Pool closely guided the students together with technical study tutors Gerrie Hobbelman, Bige Tuncer and Guus Westgeest. Throughout the semester a series of exclusive workshops was given by guest lecturers including Jeroen Coenders, Jerome Decock, Lukas Feireiss, Ulrika Karlsson, Chris Speed and Patrick Teuffel. Students could also broaden their knowledge and skills in the Literature and Media course taught by Bert Bongers and Research and Design course led by Hans Hubers who also coordinates the Hyperbody education. Each of the students was encouraged to be creative and innovative; they were all allowed to develop even the craziest concepts as long as their projects obeyed the following design rules of the master plan:

1. Functional program is a mixed use development of 1.000.000m^3 of built-up area
2. It is contained in a virtual sphere of 8000000m^3, leaving 75% open space for bringing light into the large urban body
3. Location will be right in the heart of the 751 site
4. Students are provided with a 3d puzzle of as many interlocking parts as there are students.
5. Each of such 3d zones communicates and negotiates only with its immediate neighbours.
6. Each piece of the 3d puzzle has a specific program of requirements [housing, offices, commercial, cultural, educational, leisure]
7. Each zone administrates their data input, data processing and data output and communicates the parameters in a dynamic database with their immediate neighbours
8. Each zone has to structurally support itself and communicates data of structural loads with their immediate neighbours
9. The sphere must produce as much energy as it consumes

The main logic behind setting up those rules was formed based on the idea of swarm behaviour. As prof. Kas Oosterhuis defines it, in a swarm each of its parts exclusively communicates with its direct neighbours. The global shape of the swarm is not imposed by any of these swarming members. There is no leader and elements of the swarm don't need to have an awareness of the whole. The outcome of the swarm is a balanced result of emergent bi-directional interactions between its acting members and of top-down external conditions. In the 751 master plan student designers were the bottom-up communicating swarm members and the tutors represented the top-down control.

Three-dimensional plot distribution according to the master planning scheme meant that all designs had to be interrelated. Some projects had to be structurally supported by other ones; some other had to provide access to those that were separated from the external environment. All projects had to work together to manage sunlight access, people, water, energy and waste circulation and many other problems.

In this setup students could not design without consideration to how designs of their neighbours were unfolding. Every

在这样的设定当中，学生们如果不考虑相邻部分的设计进展则无法完成设计。每一个特定的设计决策可能潜在地、直接地或者间接地影响接下来的设计决策，不仅是特定的设计者的，甚至是整个设计组的决策。在设计过程中，学生们认识到在这样的条件下预设的和固定的设计方法变得十分无效。周围情况迫使他们以弹性的和参数化的方式进行思考工作，因此，当它们的环境不断变化的时候，设计能够实时改变。参数化的设计规则的原理或许很容易，假设设计者能够很快地使用数字化时代带给我们的所有技术。

然而，这对那些具有传统的教育背景，缺乏需要的技术技能的人仍然比较困难。因此学生们学习了介绍参数化三维生成设计以及利用Virtools Dev以产生实时三维互动设计的课程。他们分为三个技术学习小组，以帮助他们解决在设计任务当中必将碰到的技术困难。其中一个小组负责组织所有进行的设计之间的数据交换；第二个小组负责所有设计方案的结构可行性；第三个小组研究生态问题，主要关注能量循环和确保整个球当中的能量消耗和生产相平衡。

数据交换小组面临的压力最大，因为他们负责发展23个设计者之间信息交换的手段。这需要立刻在设计过程中得到采用。为了这一目的，学生与他们的技术和设计教师发展了一个数据库的原型以储存和交换设计数据和交互式的三维工具，以确定信息能够在球体内部得到交换。这些交换的信息可以是结构负荷、运输和其他方面定义的数据。

不断交换的数据已经作为实时的参数被共同采纳。每一设计基地预先设定了边界，这意味着影响了每一方案的周围的所有信息也与表面的位置有关。结构荷载能够表达为锚固在表面的三维向量。人流和车流将只是一个表面的正或负值。这一信息能够在数据库中容易地建立并且立即被把它作为方案参数的使用者提取。而且，在这一框架中，更多的数据可以容易地添加。

设计系统、技术和方法与设计进程同时发展。因为这点，设计方案并不是到了设计的最后阶段才拼接为一个整体。每人都在等待那一刻，我们所有人都感到惊讶，当我们最终体验这一所有方案组成的虚拟模型当中的三维城市，所有的设计者形成一个庞大的复合结构。然而，另一方面，每一位学生设计者作为一个独立的设计概念都有绝对的唯一性。它们当中的一些表达为定义清晰、固定的建筑空间。其他一些根据居住者的需求而弹性变化。许多方案是由大量细胞元素组成，同时其他一些只是一个单一形体，在一个表皮之下包含了所有内部空间。它们同其他部分一起填充了整个球体。

这本书的主要部分是收集了那些方案的摘要。那些工作可以看成不同设计和想法的集合。然而，更本质地，它可以被看作未来城市的新图像。一座三维城市，可以与我们的生活互动，能够包含大量的新技术。这一切将以我们以前从未想象过的方式展开。

* Hyperbody是代尔伏特理工大学Kas Oosterhuis教授指导的研究所。Hyperbody的目标是研究互动建筑以及发展其实际应用。

particular design decision could potentially, directly or indirectly, influence all following design decisions, not only of the particular designer, but of the entire team. During the design process students have realised themselves that under such conditions designing in a pre-defined, fixed way becomes very inefficient. The circumstances have forced them to think and work flexibly and parametrically, so that projects could be changed instantly while their surroundings were continuously evolving.

Parametric design may be easy in its principle, assuming that designers can swiftly use all technologies that the digital era has brought to us. However, it might have still been difficult for those with a traditional, top-down educational background and lack of needed technical skills. Therefore students were given introductory courses to Bentley Generative Components for parametric 3d form designing and Virtools Dev to create programs that generate real-time 3d interaction. They were also split into three technical study groups to help them with particular technical difficulties which they were bound to encounter in their assignment. One of the groups was responsible for organising the data exchange between all evolving designs. The second group was taking care of structural feasibility of all projects and the third group has been researching ecological issues, mostly focused on energy circulation and made sure that the global energy consumption and production of all projects together would be balanced.

The biggest pressure has been put on the data exchange group, as they were the ones responsible for developing means of exchanging information between the 23 designs. It needed to be immediately applied in the design process. For this purpose students together with their technical and design tutors developed a database prototype for storing and exchanging design data and an interactive 3d tool to locate points on which the information could be exchanged within the sphere. This exchanged information was related to structural loads, transportation and other, custom defined data. The continuously exchanged data had been acting as instant parameters for mutual adaptation. Each design plot had predefined boundaries. This meant that all information potentially affecting the neighbours of each of the projects was also related to a position on that surface. Structural loads could be expressed as three dimensional vectors anchored on that surface. Flow of people or cars would just be a positive or negative value on that surface. This information could be easily structured in a database and immediately accessed by parties that were using it as parameters of their projects. Moreover, in such framework, more parameters could easily be added.

The design system, techniques and methods were being developed in parallel to the design process. For this, it was not until the very last phase of the design process that projects could be assembled into one whole and the total outcome might have been verified. Everyone was waiting for that moment and we were all astonished when we were ultimately able to navigate through the three dimensional city in the virtual model assembled out of all projects. All designs were forming one, enormous complex structure. However, on the other hand, each of the student designs has been absolutely unique as a standalone concept. Some of them were embodying well defined, fixed architectural spaces. Other ones were flexibly responding to demands of their inhabitants. Many of them consisted of a high number of cellular elements, while others were just singular bodies embodying all inner spaces under one skin. They all filled up the volume of the sphere with respect to each other.

Main part of this book is the collection of summaries of those projects. That work can be considered as a collection of diverse designs and ideas, yet more essentially; it has to be seen as a new vision for the city of the future. A city that would grow in three dimensions, which would interact with our lives and which would embody numerous new technologies. All this in ways that we could have never imagined before.

* Hyperbody is a group at Delft University of Technology directed by prof. Kas Oosterhuis. The aim of Hyperbody is to study interactivity in architecture and to develop its practical applications.

Structural Design Study
结构设计研究
Gerrie Hobbelman

The students were divided in three groups. Each of those groups had a different specialization. One of them was structural design.

Structural design students were faced with a very difficult task. The design zones were stacked one over another. Therefore, all students had to make sure that on one hand there is enough structural support for their projects, on the other hand they had to provide support for others that were located above or around them.

Before the students started working, they had to develop an understanding of three dimensional structures and how the forces are transmitted in them. One of the methods that help them with this was to build a human pyramid and personally feel the forces on themselves coming from others they were carrying. This was not done in reality, it would have been too dangerous with 10 persons. The pyramid was put in a picture and every student adopted one person in the pyramid and described the forces he or she had to carry. The person on top had the easiest part. Persons in the basis of the pyramid had to carry themselves plus the others standing above. This gave them a good insight in the matter of stacking structures upon each other.

In the process of their work they had to identify points on the borders of their zones where forces were transmitted, the value of those forces and the direction in which the transmission was occurring. Once the model of all forces in the entire sphere was made, the whole project could have been statically calculated.

This task would be already complex enough if the designs were fixed. However, during the entire design process all designs were in constant development. Throughout the whole semester

学生们分为三组。每组有着不同的专门方向，其中一组就是结构设计。

结构设计学生面对着一个非常困难的任务。设计分区一个个堆在一起。这样，所有学生都要保证一方面有足够的结构支撑其项目本身，另一方面，他们必须为其上面和周围的地块提供支撑。

在学生们开始设计之前，他们必须建立起对三维结构的理解，以及其中的力是怎样传输的。帮助他们理解的一种方法就是建立一个人员金字塔，并亲自体验他们承受的来自别人的力。这不是在现实中做的，因为这样做对于10个人来说太危险。金字塔画在图上，每个学生认领金字塔上的一个人，并描述他或她必须承受的力。在顶端的人最轻松。位于金字塔底部的人要承受自身和上面人的重量。这让他们很好理解了这种堆砌结构的实质。

他们在工作过程中必须确定其地块边界上用来传力的点的位置、这些力的大小，以及传力的方向。整个球体所有力的模型一旦建立，那么整个项目就可以进行静力计算了。

即使设计已确定，这项任务也是相当复杂的。然而，在整个设计过程中，所有设计都在不断发展变化。在整个学期里，新想法不断涌现，项目只能不断重构。这样，为确保整个结构的稳定，即使每次最微小的变化也要进行重新计算。在这一点上，该组与数据交换技术组合作建立了一个系统，可根据所有的变化进行信息动态交换，并在结构模型上得到更新。最终，建立起了一个弹性计算系统，在每次数据更新后自动对整个模型进行重新计算。结构设计组学生最先将项目模型一起合并为一个文件，以验证所有设计连接起来后在静力上能否建造这个巨型结构。

还可以发现，并非每名学生都能完全明白其在结构上的责任。设计球体底层的学生必须承受其上方很多地块。但是他们却仍然选择设计那种几乎连自己都承受不住的结构，根本不管其他地块。一个学生设计球体非常中心的地块，是作为核心而从头开始建造的理想位置，但她却决定设计一个用聚合墙体建造的很高的水池，毫无结构支撑作用。

new ideas kept on emerging and projects were subject to constant reconfiguration. Thus, in order to make sure that the whole structure was stable; it would have to be recalculated every time when even the slightest change was made. In this point the group cooperated with data exchange technical group to create a system that would allow dynamic exchange of information upon all changes and updates in the structural model. Eventually a flexible calculation system could have potentially been established in order to automatically recalculate the whole model after every data update. Structural design group students were the first ones to put the project models together in one file to verify if it was statically possible to build the emerging mega structure of all connected designs.

Another observation was that not every student was fully aware of his responsibility in a structural point of view. Students with a plot lower in the sphere had to carry many other plots above. Still they choose for a design which could hardly carry itself, let alone a set of plots extra. One student had a plot in the very center of the sphere that went up from the ground all the way up, an ideal situation for a core construction, but she decided to make a tall water basin with polymer walls that gave no structural support at all.

Data Exchange Study
数据交换研究

Bige Tunçer

In collaborative design, communication among the participants is crucial for the success of the process and the resulting product. In the E-motive architecture design studio, the design brief requires the students to intensely collaborate. It is important to apply a methodology for effective communication in such a studio. In this context, the task of the data exchange group was to ensure that the project data and information that need to be communicated among individuals and groups is structured, stored, managed and presented in a high-quality manner. The data exchange group defined means and mechanisms in order to enable the formal communication processes among the designers of the 23 interlocking 3D plots in the bounding sphere.

The data exchange group first developed a data model for the purpose of all design studio students to negotiate and establish connections between their plots (Figure 1). The goal of the data model is to make sure that the all relevant data objects required for communication within the collaborative design process are completely and accurately represented.

After the definition of the data model, the data exchange group proceeded to the design of the common database. The goal of the database is to store information about connections between plots for the flow of loads, energy, light, etc. and to record certain properties of each plot which need to be balanced according to the program requirements (developed volume, functions, etc.). Conceptually, the database is designed as a tree (Figure 3).

The database was implemented using MySQL, which is the most popular Open Source relational SQL database. The database consists of five tables:

在合作设计中，参与者之间的沟通对于过程和结果的成功是至关重要的。在E-motive建筑设计工作室中，设计大纲要求学生们密切合作。在这样一个工作室中，运用有效方法进行沟通很重要。在此情况下，数据交换组的任务是，确保需要在个人之间及小组之间相互沟通的项目数据和信息能够建立、存储、管理，并进行高品质表达。数据交换组定义了相应方法和机制，以使球体范围内的23个连锁三维地块的设计者之间能够进行有效沟通。

为了让设计工作室所有学生能够商讨并在其地块之间建立连接，数据交换组首先建立了一个数据模型（图1）。数据模型的目标是，确保合作设计过程中需要沟通的所有相关数据对象都能完整而准确地表达。

定义了数据模型之后，数据交换组继续进行公共数据库的设计。该数据库的目标是存储有关地块间荷载、能量和光线等流动的连接信息，记录每个地块需要根据功能需求（开发体量、功能等）进行平衡的特定属性。从概念上说，该数据库被设计为树形结构（图3）。

该数据库是用MySQL来实现的，这是源代码开放的最流行的SQL关系数据库。该数据库由五个工作表组成：

——用户：该工作表记录参与学生的所有相关信息，如姓名、地块编号和电子邮箱地址等。

——功能：该工作表记录每个地块的功能（以立方米为单位、包括人员数量）。整个球体不同功能的总和也可用它来计算。

——表面：该工作表记录属于每个地块的所有表面。表面工作表包含每个给定地块表面的局部坐标系。该坐标系用的是坐标偶，而不是用三维坐标来为每个表面确定法线，定义连接，不必定义表面的边界，并允许任何与一个表面相关联的Z值，以指示流速（这很重要，因为很多连接可能随时间有强度变化）。数据交换组为确认不同表面建立了一种特定的命名

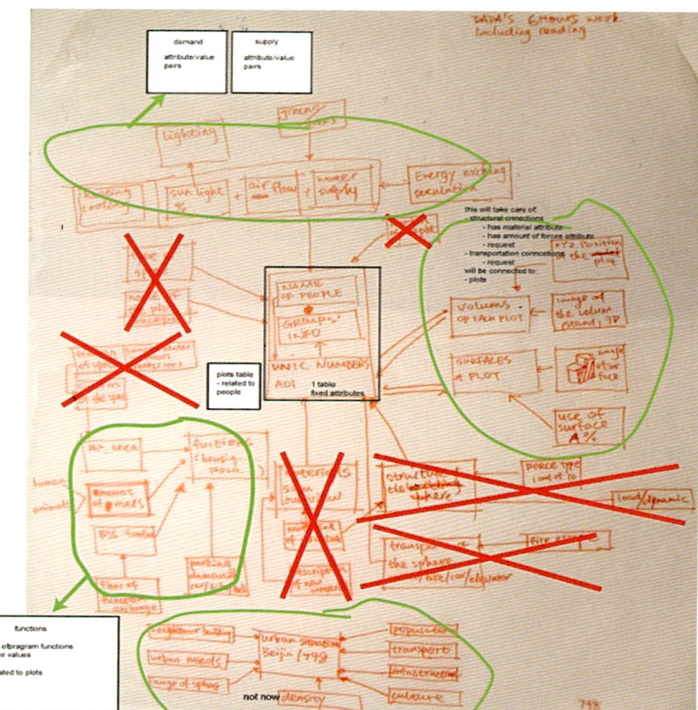

figure 1. Data model sketch

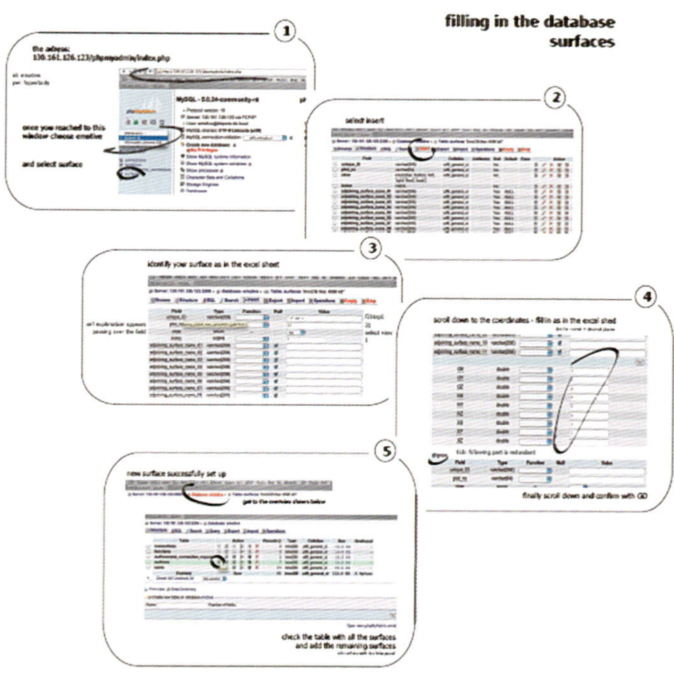

figure 2. Tutorial for students to fill in the database

-Users: This table records all relevant information about the involved students such as name, plot number and e-mail address.

-Functions: This table records the functions (in m³ and the amount of people) of each plot. Calculations summing various functions for the whole sphere might be applied to it.

-Surfaces: This table records all surfaces belonging to each plot. The surfaces table contains the local coordinate system for each surface of a given plot. These local coordinate systems allow ordered pairs, instead of 3D coordinates to be used to define connections, create normals for each surface without having to define the border of the surface, and allow any Z values associated with a surface to indicate flow rate (this is important as many connections may vary in intensity over time). A specific nomenclature was developed by the data exchange group for the identification of the surfaces. The surfaces that are shared by plots are identified and numbered with a specific numbering system in order to determine neighbourhood relationships and neighbouring surfaces.

方法。不同地块共享的表面用一种特定的编号系统来确认和编号，以确定相邻关系和相邻表面。

——表面_连接_请求：如果相邻地块想建立连接，那么其中之一要用该工作表发出请求。数据库中的一个用户不必定义很多参数就能向另一用户发出连接请求。

——连接：连接请求一旦得到确认，就会显示于该表格上。该表格还用来列出连接点，并处理结构、能量和其他信息及地块间通过这些连接点的基础设施交换。连接范围很广，可以是任何物理或象征物，或者能量。

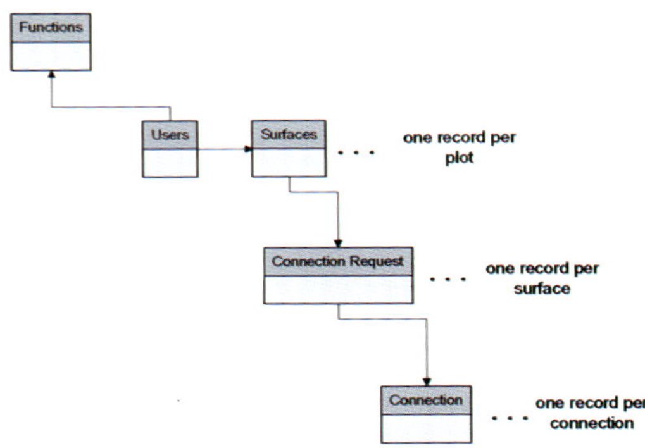

Figure 3. Conceptual Database Model

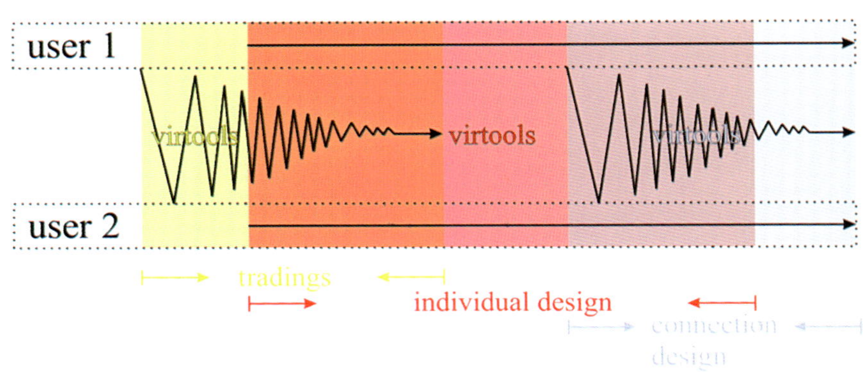

Figure 4 . Ideal situation: All phases of design realized in Virtools interface with permanent interaction through this interface.

因为学生们要自己负责用所有信息来填充数据库，所以数据交换组开发了一个如何填表的指南。为尽可能保持简单，决定用抓屏加简短说明的方法，并标记重要按钮。

为了给该数据库用户提供直观简单的使用界面，数据库连接必须完美地整合于设计环境中，主要是Virtools环境中（图4）。这就让所有用户都能立即看到数据库中的所有变化。因为所有变化结果都是图示表达地，所以这是个对用户更加友好的合作设计环境，可以更加高效和有用地交流。该目标在此课程中得到了相当程度地实现。

为学生们建立了第一个通用的Virtools模型，以鉴别和命名其地块表面（图5）。

在该工作室的时间框架内，曾设计了Virtools与数据库的界面，但未实施（图6）。该界面包括观察者（待建立）、一个数据库连接脚本和几个工具。下面说说三个工具。

一个是连接工具（十字），在一个表面上建立连接。该工具填充了数据库中的所有适当位置。连接工具一旦使用，断面工具就可用来绘制连接的断面图，填充数据库更大的范围。

一个是数据库图示工具，是一个类似Excel的弹出界面，让数据库以概念化的树形来图示。这是通过使用绘图工具或者通过地块/表面目录的漫游来启动的。

一个是坐标转换工具，当处理单个地块上的表面时，该工具对于局部坐标系与全局坐标系之间的简单转换是很有用的。

在合作设计过程中，人们得到了一个易于使用的数据库，一个直观的三维界面，这将对于合作的质量、甚至在概念设计阶段就形成很有价值的贡献。

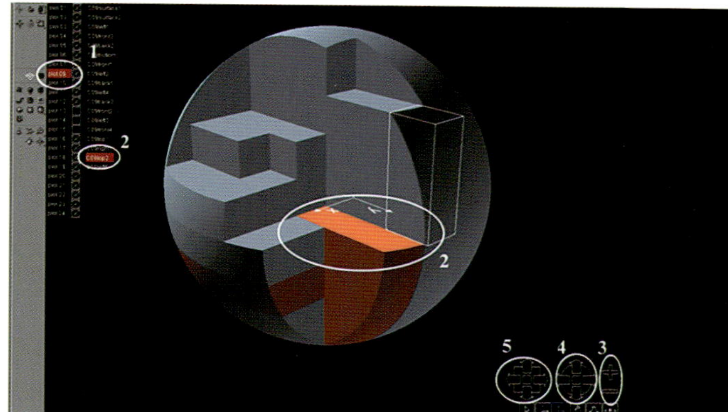
Figure 5. Virtools interface ready for integrating student designs.

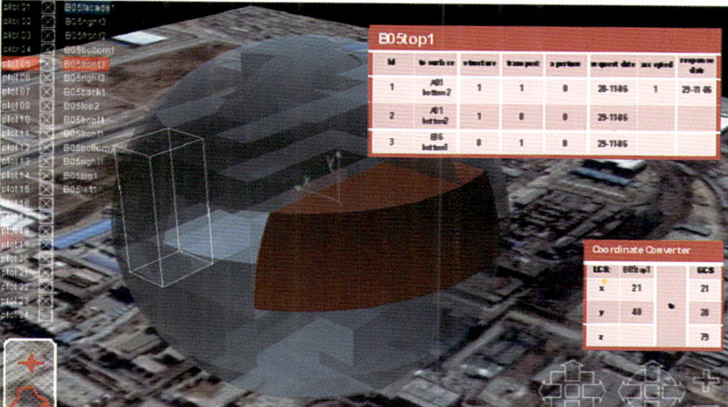
Figure 6. Mockup of proposed interface tools

-Surface_connection_request: If two neighbours want to establish a connection, one of them requests the connection using this table. This allows one user of the database to propose a connection to another user without having to define a large number of parameters.

-Connections: If the requested connection is confirmed, it appears in this table. This table is also used to specify the connection points and do the structural, energy and other information and infrastructural exchange through these connection points between the plots. Connections are broad and may refer to any exchange of physical or symbolic material or energy.

Since students themselves were responsible for filling the database with all the information, the data exchange group developed a tutorial about how to fill the various tables of the database (Figure 2). In order to keep it as simple as possible it was decided to capture screen-shots adding short instructions and marking important buttons.

In order to provide users of this database with an intuitive and easy to use interface, the database connection must ideally be integrated in the design environment, which is mainly Virtools (Figure 4). This allows all the changes in the database to be immediately visible to all users. Because results of all the changes are also graphically visualized, the collaborative design environment becomes more user-friendly and thus allows a more efficient and effective communication. This goal was realized in this course to a certain extent.

The first common Virtools model was developed in order for the students to identify and name the surfaces of their plots (Figure 5).

Within the time frame of this studio, the Virtools interface to the database was designed, but not implemented (Figure 6). This interface consists of the viewer (which was created), a database connection script, and several tools. Three tools are described below.

A connection tool (the cross) places a connection on a surface. This tool also fills in all appropriate fields in the database. Once the connection tool is used the cross-section tool can be used to draw the cross-section of a connection, filling in more fields of the database.

A database visualization tool is a pop-up Excel-like interface that allows the database to be visualized in its conceptual tree form. This is activated by the use of the drawing tools or by navigating through the plot/surface tree.

A coordinate converter is highly useful to easily convert between local coordinate systems and the global coordinate system when dealing with surfaces on individual plots.

An easy to use database with an intuitive 3D interface available early on in this collaborative design process will form a valuable contribution to the quality of the collaboration, even in the conceptual design phase.

Ecology and Energy Study
生态和能量研究

Guus Westgeest

前言

生态是建成环境可持续发展的一部分。Hyperbody MSc3课程的学生们被要求在其设计中进行此项研究。这个学期的任务是在北京751基地上建造一个巨型复合建筑。从生态学观点出发，所有学生必须遵循一个规则：能量产生于设计基地，这个复合建筑产出的能量要至少与其消耗的相同。整个开发必须产出能量以自用。能量使用和生产可在地块间交易。在选择能量生产和消耗方法时要充分考虑可持续问题。

生态与可持续发展

“可持续发展”的概念（根据Brundtland,1987）虽得到了广泛支持，但并未带来所需的所有行动。似乎在任何行动进行之前，有必要将可持续性分为特定的主题。最为熟悉的步骤是将可持续性划分为一些独立的课题：社会可持续发展、生态可持续发展和经济可持续发展（人＋地球＋收益，根据1992年里约热内卢全球首脑会议，和Elkington，1998的定义）。在荷兰的实践中，可持续发展被简单限制为一个主题，诸如居住环境的健康或能量效率。

建成环境的可持续发展

从事可持续发展工作可有很多形式：诸如帮助本土居民、帮助非正式经济、绿色发展和庇护穷困者。所有这些方面都是与建成环境有关的。那么作为建筑师和城市规划师从哪里出发呢？其实应当首先强调自身专业。对建成环境来说，这是空间品质问题，是项目的内在品质。如果这是设计者的目标，那么可持续发展的主题就会成为他或者她的方法。所以，通过将可持续三角转变为四面体（Duijvestein，2003），建成环境的可持续发展便可定义。这就产生了图1。人们需要从一个特定的利益区间开始，通过定义这样一个主题，该模型指出了可持续发展的一个重要方面。所以该四面体也可用于其他目标。

可持续建筑模型的功能

没有什么完美的模型可采用，模型是与其目的相关的，模

Introduction

Ecology is a part of Sustainable Development in the built environment. Students of the Hyperbody in the MSc3 course had to implement the requirement of this study in their designs. In this semester the assignment was an enormous hybrid building in Beijing on the 751 site. From the ecological point of view all students had to obey one rule: energy is produced on the design site and the hybrid building produces at least the same amount of energy as it consumes. The whole development has to produce energy for its own use. Energy use and production could be traded between the plots. Sustainability issues needed to be addressed when choosing energy production and consumption methods.

Ecology in relation to Sustainable Development

The concept of “sustainable development” (according to Brundtland, 1987) enjoys widespread support, but it does not lead to all the needed actions. It seems to be necessary to divide sustainability up into particular themes before any action can be taken. The most familiar step is breaking sustainability into separate topics: social sustainable development, ecological sustainable development and economic sustainable development (People + Planet + Profit, according to the Rio de Janeiro Earth Summit 1992 and defined by Elkington, 1998). In Dutch practice sustainable development is easily limited to one theme like health or energy efficiency in the living environment.

Sustainable Development in the Built Environment

Working on a sustainable development can have many forms; like helping out indigenous people, helping out informal economies, green development and giving shelter to the poor. All aspects are related to the built environment. So where to start as an architect or a city planner? In practice it's easy to start off with

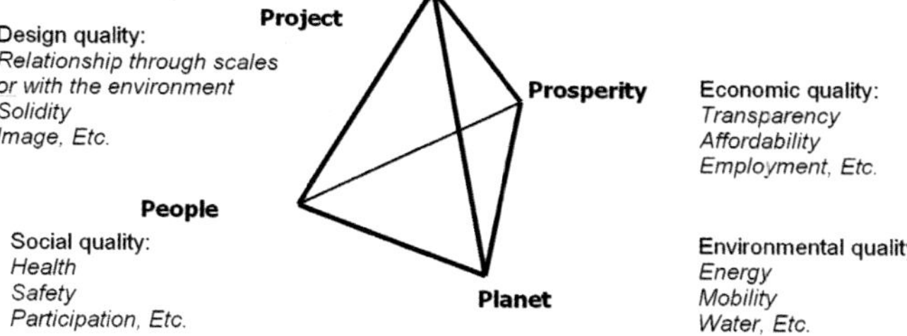

Figure 1. Sustainable design model

型是个沟通工具。3-P三角联系着可持续发展的中心主题。模型还可实现设计过程中的一个角色，它们能够帮助作出决策。设计模型在于简化和解释任务，模型还可实现对现实的评估。城市究竟有多么可持续？评估模型是计算和比较现存环境品质的一个睿智方法。用LCA来描述材料的可持续性数值，这就是一个评估模型的例子。沟通和设计模型的一个例子就是DCBA方法（D：正常状态，C：正确的正常使用，B：对环境的最小化破坏，A：理想状态）。

751项目的生态系统

北京遭受着工业和交通带来的严重污染和恶劣的空气质量。中国北部和西北部沙漠的侵蚀带来的沙尘，导致了让这个城市苦恼不堪的季节性沙尘暴。单是2006年初的4个月就有不少于8次的这种风暴。为准备2008年夏季奥运会，北京近来已进行努力来净化城市。根据政府消息，中国在过去6年中已采取措施，努力增强草地生态，来抑止沙暴对国家首都北京和港口城市天津的影响。感谢这个项目的实施，在过去6年内沙化区域已减少15.5万hm^2。

798基地的生态分析

在城市水体系统优化方面，玉渊潭，南壕河及通惠河－高碑店湖的水体系统已进行疏浚，并变为观光滨水走廊。每日城市污水处理能力达到了180万吨，比去年提高了25.9%，年污水处理率达到45%。已在250个居住区实行垃圾引导分类和收集项目。高安屯垃圾处理厂首期已在2002年投产。城市垃圾每天都得到清扫和处理。2002年有大约86.5%的城市垃圾作了环境无害化处理。

有4681个烧煤锅炉已变为使用清洁能源。北京的天然气供应达到20.5亿m^3。超过400万平方米的供暖空间现在已用电来作燃料。北京有大约80.4%的车辆达到了环境排放标准。北京还

emphasising your own profession. For the built environment this is the spatial quality, the intrinsic quality of a project. If this is a designer's goal, the themes of sustainable development can be his or her means. So a sustainable development in thezt built environment can be defined in turns the sustainability triangle into a tetrahedron (Duijvestein, 2003). This leads to the model of figure 1. The model points out an important aspect in sustainable development; in defining the topic one needs a specific field of interest to start from. So the tetrahedron can also be used for other objectives.

The function of a model in Sustainable Building

There isn't the perfect model to be drawn. Models are related to their purpose. A model is a communication tool. The triangle of the triple-p communicates the central themes of a sustainable development. Models can also fulfill a role in the design process; they can help in making decisions. Design models are there for simplifying and explaining the tasks. Models can also fulfill a role in evaluating the reality. How sustainable is this city? An evaluating model is a smart way to calculate and to compare the qualities of an existing environment. An example of an evaluating model is a LCA describing the sustainability score of a material. An example of a communication and design model is the DCBA method. [D: The normal situation, C: Correct normal use, B: Minimizing damage to the environment, A: The most favorable or ideal situation]

Ecology in the 751 project

Beijing suffers from heavy pollution and poor air quality from industry and traffic. Dust from erosion of deserts in northern and northwestern China result in seasonal dust storms that plague the city. In the first four months of 2006 alone, there were no fewer than eight such storms. Efforts have been made of late to clean up Beijing in preparation for the 2008 Summer Olympics. China's efforts to improve grassland ecology for the purpose of curbing sandstorms affecting the national capital Beijing and port city of Tianjin have paid off in the past six years, according to government sources. Thanks to the implementation of the project, the sandy areas in the project have been reduced by 155,000 hectares in the past six years.

Beijing on a day after rain and a sunny but polluted day.

DCBA design model for the 798 factory site

在努力建立城市绿化带。2002年，在城市内建立了超过51个绿化区，在城市周边建立了超过110km²的绿化带。在北京总共有176.5万辆汽车，其中私家车45.8万辆。总共592条公交线路每年可提供49.2亿人次的载运能力。北京国际机场，凭借其新近开放的站厅，已成为亚洲最繁忙的空中交通中心之一。

从生态观点出发的规则设定

能量：

——球体作为一个最终产品，在球体的南部为我们提供了一个巨大区域进行太阳能收集。

——在球体内部废弃能量的再利用。

——针对球体的整体解决方案（参照德国议会大厦穹顶的自然通风方法）。

建筑材料：

——使用地方材料。

——100%使用可循环利用的材料。

结构：

——“可拆解”，这样当球体生命结束时，无须特别努力就可很容易将其移除。

水：

——使用雨水作为厕所和清洗用水的总体解决方案。

垃圾：

——球体产生的每件垃圾都应可循环并转化为能量。球体不向外界释放任何垃圾。

交通：

——仅在球体底部可使用小汽车。

——球体内的交通将是生态优先的。

The Ecological analysis of the 798 site

In terms of urban water system improvement, the Yuyuantan, South Moat and Tonghui River-Gaobeidian Lake water system have been dredged and were turned into a site-seeing waterway route. Daily urban sewage processing capability reached 1.8 million tons, which is 25.9% higher than that of the previous year. Annual sewage processing rate equaled 45%. Pilot garbage sorting and collection projects have been carried out in 250 inhabitant areas. The first phase of Gao'antun Garbage Handling Factory became operational in 2002. Urban garbage has been cleaned and processed daily. About 86.5% of the urban garbage in 2002 was processed to be environmentally harmless.

There are 4,681 coal-consuming furnaces that have switched its power supply from coal to clean energies. Natural gas supply in Beijing reached 2.05 billion cubic meters. Over four million square meters of heating space are now fuelled by electricity. A total of 1,800 public transportation buses are now using clean fuel. About 80.4% of all the vehicles in Beijing have met the environmental standard of exhaust emission. Beijing is also working on the establishment of greenbelt in the city. In 2002, over 51 green areas in urban region and 110 square kilometers of greenbelts around the city were built. Road density was 85.4 kilometers every 100 square kilometers. There are altogether 1.765 million vehicles in Beijing, with 458,000 private ones. Beijing also has advanced public transportation systems. A total of 592 public bus routes can offer an annual transportation capacity of 4.92 billion people/times. Beijing International Airport, with its newly opened terminal, has become one of the busiest air traffic centers in Asia.

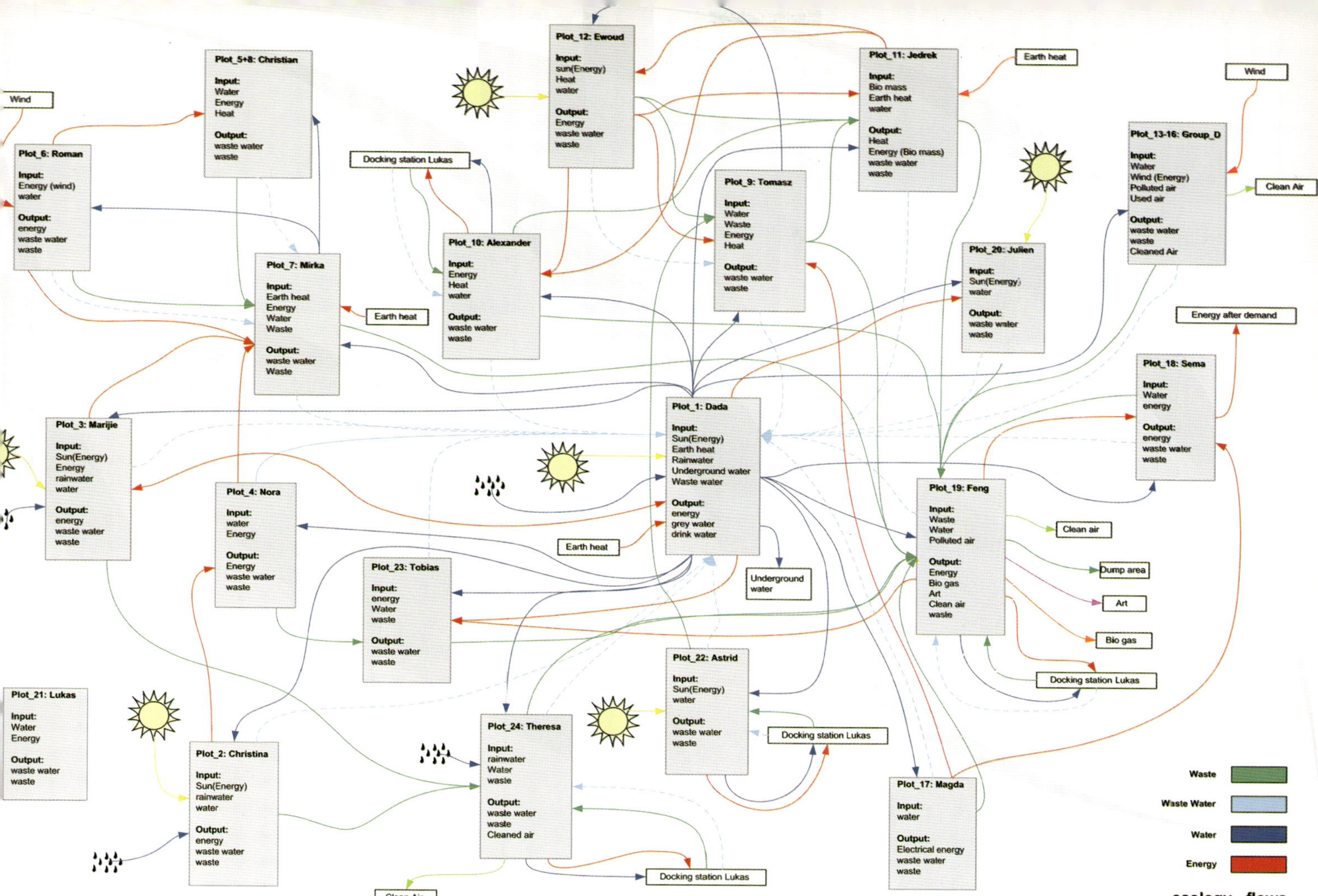

Circulation of energy in the project

Rules set for the project from the ecological point of view:

Energy:

-when concerning sphere as a final product that gives us a huge area to collect sun on southern part of the sphere.

-reuse of already emitted energy inside the sphere.

-global solution for the sphere (natural ven to work as Reichstag dome.

Building materials:

-Use of local materials

-Choosing 100% recyclable materials

Structure:

-"Demolishable" so when the sphere life will end it could be easily removed without special effort.

Water:

-Global solution for using rainwater for flushing and cleaning

Waste:

-Every waste produced in the sphere should be recycled or converted to energy. The sphere is not producing any waste to outside.

Transport:

-Car access only available on lower parts of the sphere.

-The transportation inside the sphere would be pro ecological.

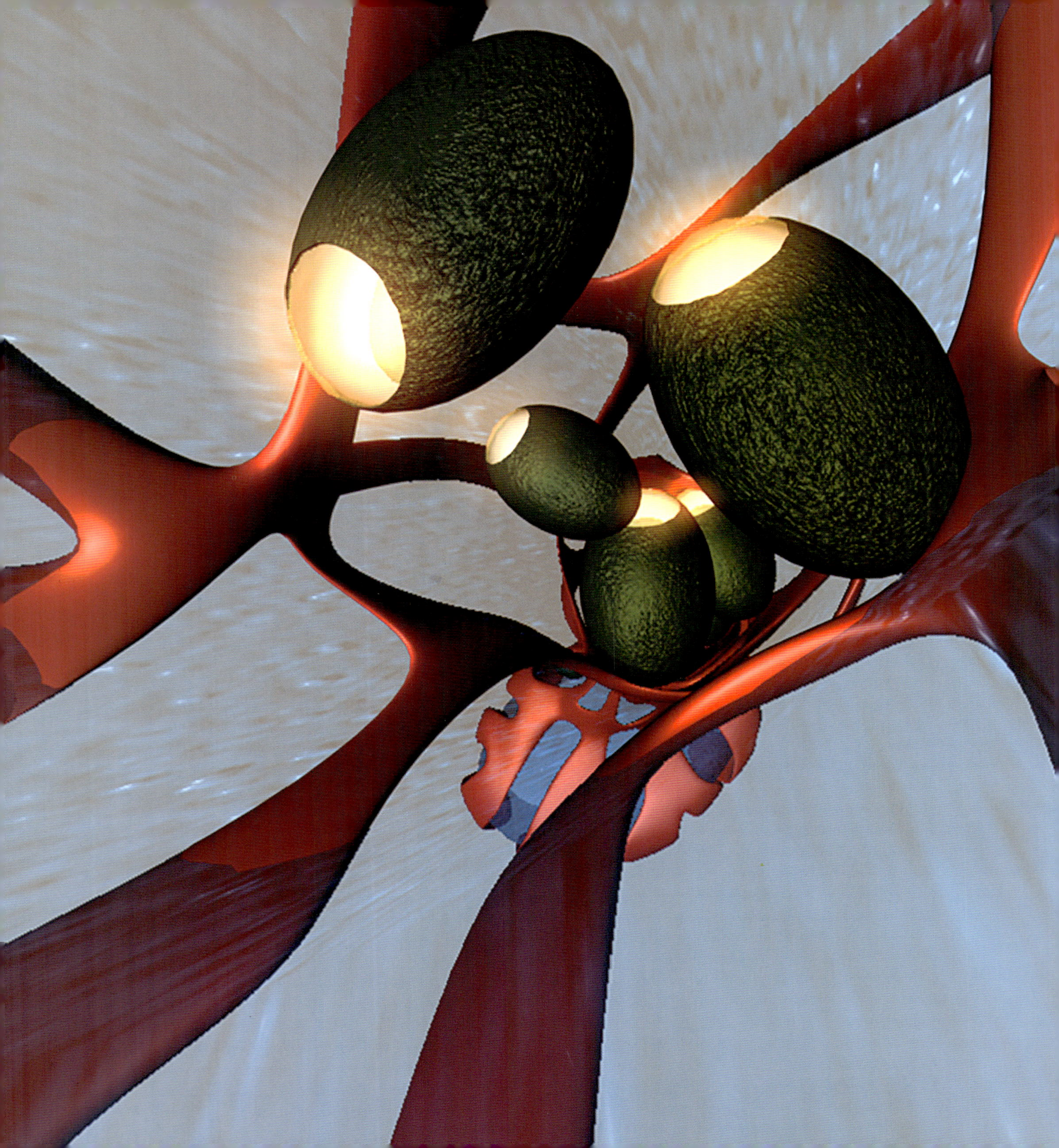

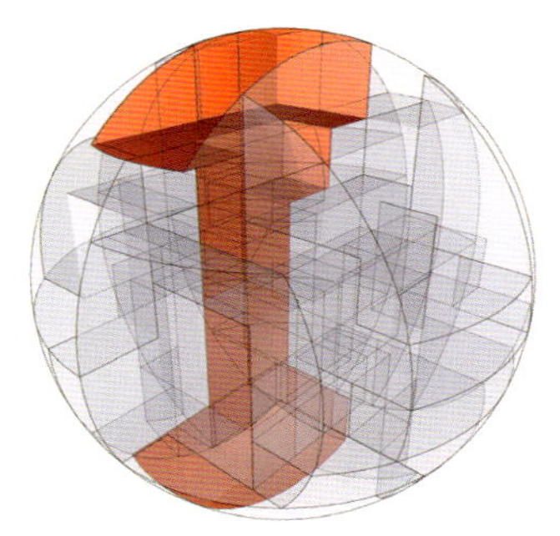

Zone 1 - The Watertower
水楼
Da Wang

在数码时代，你可以通过网络或移动通信设备来定制你所需要的空间。你可以保持静止，而你所需要的空间会向你移动。这座高163米的高层承载了90000m³的商业休闲功能空间。在拥挤和快节奏的北京，提倡一种享受悠闲缓慢的城市生活模式。而你也再也不用为迷路或堵车而烦恼。

为了实现空间的自由移动，此设计采用了“水”为媒介。去除空间的重力约束，在水中的运动没有预定的路线而由每个使用者的实时定制而产生。根据水中力的分布而采用受力最合理的泡状空间，同时也提供了一种不同寻常的空间体验。当来自这个球综合体的其他邻居需要定制空间来购物或跳舞，他们可以通过实时更新的3D网站来浏览整个水楼中的空间分布使用状况，输入所需要的时间地点，空间会自动移动到离你最近的转接站等候被使用。空闲的空间还可以根据以往的信息分析结果进行群聚行为，在最常被使用的地点周围分布待用。整个水楼有两层表皮：一是柔软有韧性的透明表皮；一是坚固稳定的钢筋混凝土支撑表皮。

作为整个球空间中的两个贯穿空间之一，水楼也提供了慢节奏的竖向交通服务，同时承担了整个群体的冷热水供应和水净化。同时作为多余能量的储备体，对整个球群体的能量循环运作起了决定性的作用。

You will be able to order the space you need by internet or mobile phone. The space will come to you instead of you driving all the way to the place. This 163 meter high tower holds 90,000m³ commercial and leisure program. It promotes a relaxed and slow motion lifestyle in the contrast to the busy and fast Beijing city life. And you will never be lost in a traffic jam.

In order to achieve the free movements of the space, the water appears as the transportation media. It avoids the limits of the gravity. There will be no fixed route. The direction of the movements will be controlled by each order. Following the force effects within the water, bubble space comes to be the best solution. It also provides unique spatial feeling to the users.

When people from the neighbors need to use the bubble space for shopping or dancing, they may enter through the most nearby docks after ordering. Through the real-time updated website, one need to input the prefer function, time and location. Then all the available options will be highlighted with detail information. The tower is running under the swarm behavior as well. The free space will follow the calculated results move to the most popular and preferred the location and get ready to be used. The whole tower has two skins: one soft, flexible and transparent skin; one concrete, fixed and solid skin.

As one of the two towers in the entire sphere, the water tower offered the vertical slow transportation possibilities. It also has the function of collecting, refreshing, heating, supplying the water to the entire sphere. By looking at the whole sphere the water tower carries lots of ecology response.

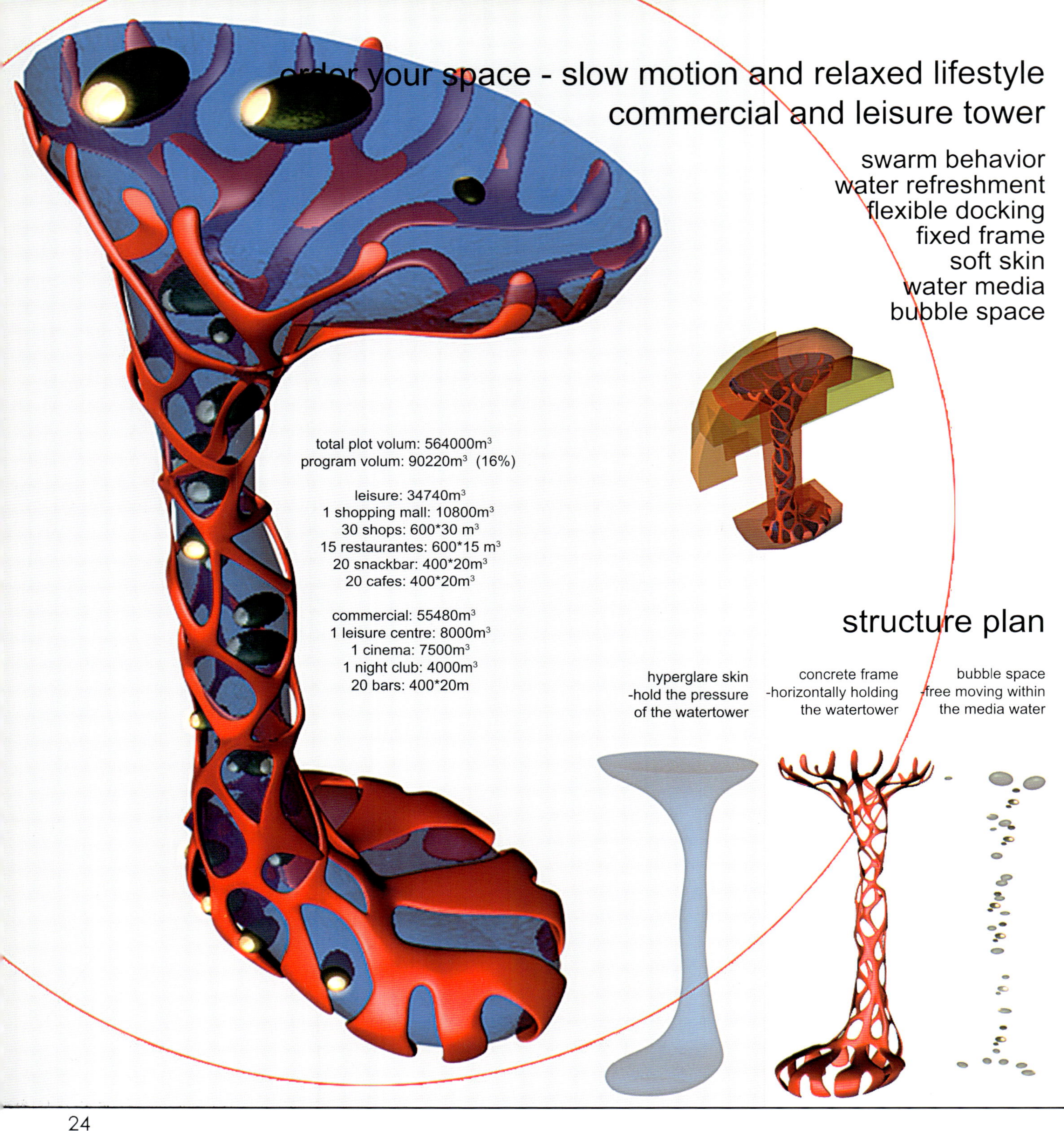
order your space - slow motion and relaxed lifestyle
commercial and leisure tower
swarm behavior
water refreshment
flexible docking
fixed frame
soft skin
water media
bubble space
total plot volum: 564000m³
program volum: 90220m³ (16%)
leisure: 34740m³
1 shopping mall: 10800m³
30 shops: 600*30 m³
15 restaurantes: 600*15 m³
20 snackbar: 400*20m³
20 cafes: 400*20m³
commercial: 55480m³
1 leisure centre: 8000m³
1 cinema: 7500m³
1 night club: 4000m³
20 bars: 400*20m
structure plan
hyperglare skin
-hold the pressure
of the watertower
concrete frame
-horizontally holding
the watertower
bubble space
-free moving within
the media water

internet interface

go through the 3d virtual tower by internet

order the time and location of the space

choose what you want to do

highlight bubbles by functions, selection or popularity level

real-time 3d model overview

drag your bubble into selection box

make combination for your orders

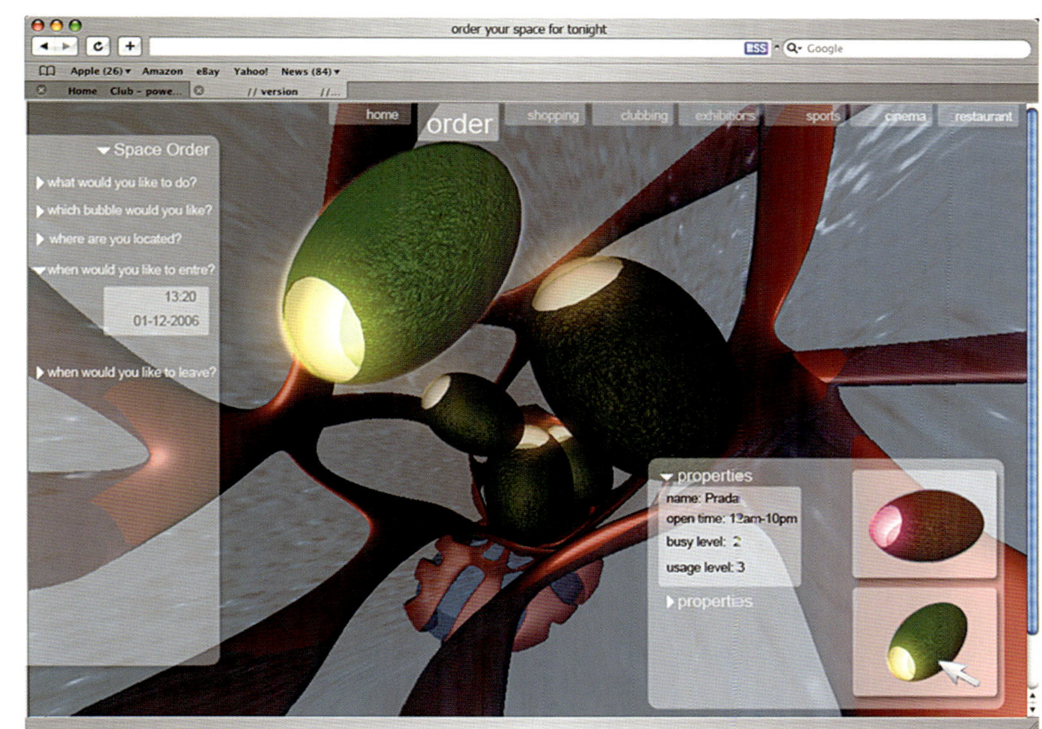

data exchange

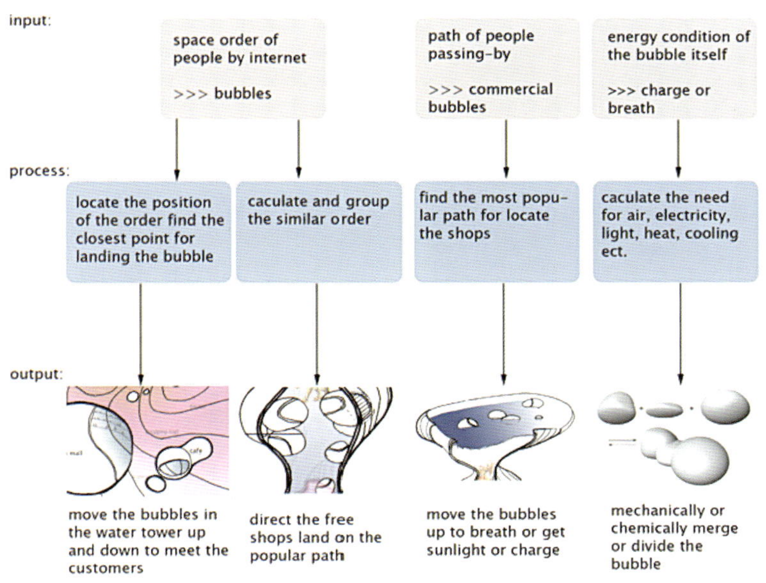

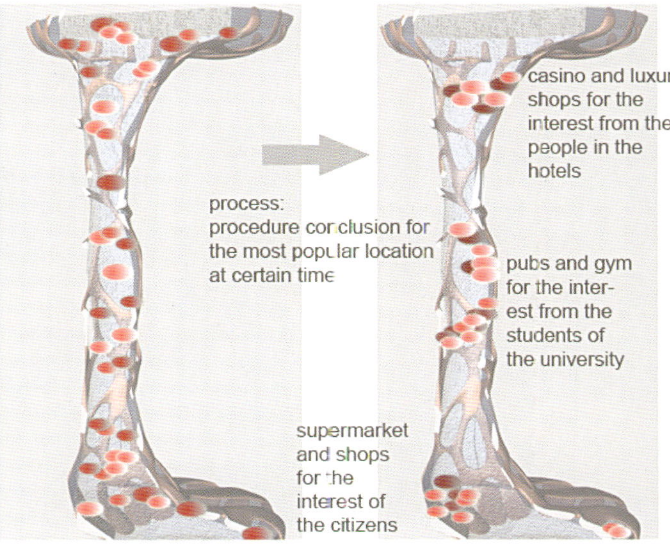

random distribution swarm distribution

ecology diagram

water storage: 89789m^3
program volum: 90220m^3

polulation of the shpere:
average day usage of water:

sun energy per day:
ground heat per day:

input:
ground heat
sun heat
rain water
waste water
electricity

output:
fresh water
hot water
middle water

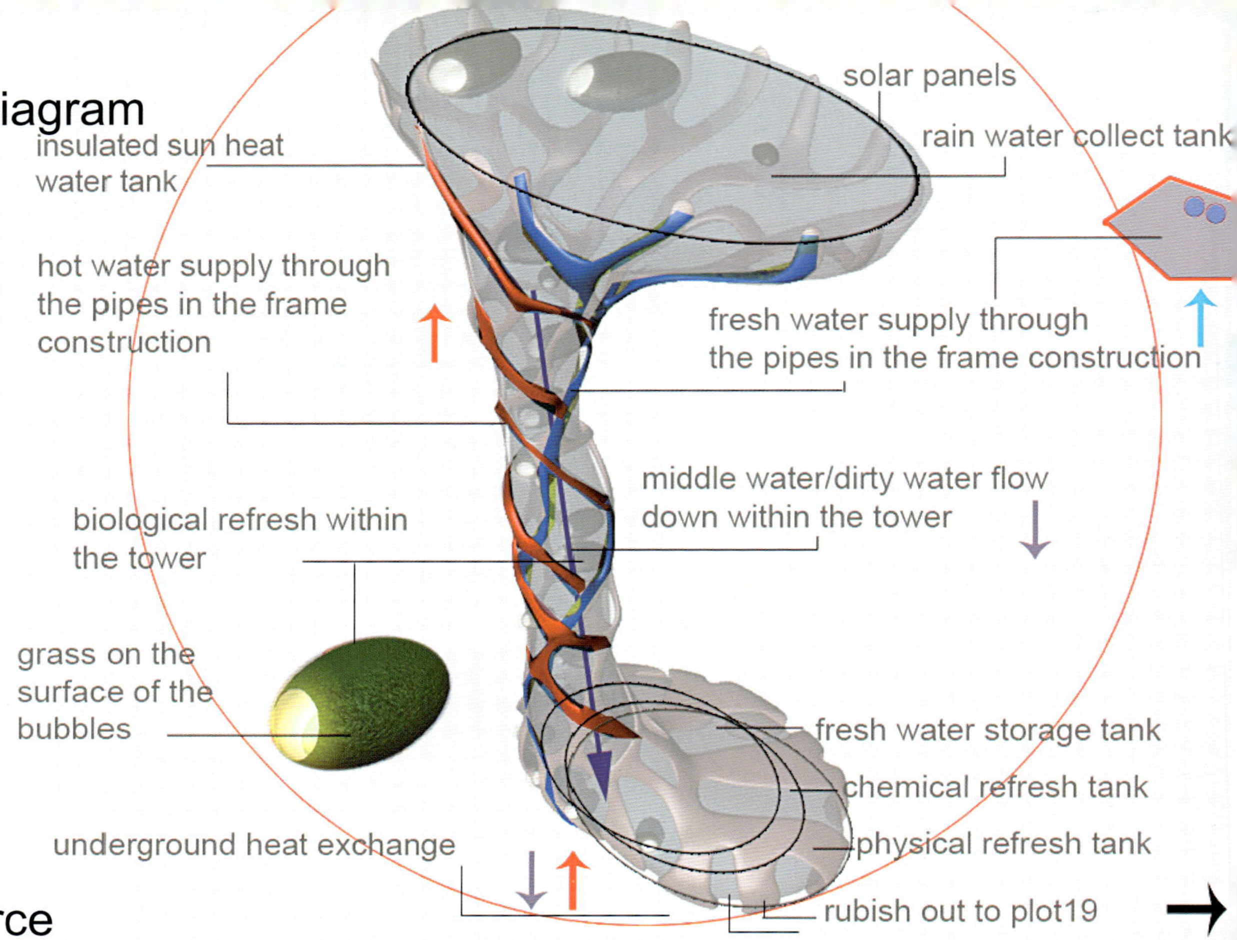

solar panels
insulated sun heat water tank
rain water collect tank
hot water supply through the pipes in the frame construction
fresh water supply through the pipes in the frame construction
middle water/dirty water flow down within the tower
biological refresh within the tower
grass on the surface of the bubbles
fresh water storage tank
chemical refresh tank
physical refresh tank
underground heat exchange
rubish out to plot19

driving force bubble structure

compressibility of water is small. an increase of ressure by 1atmosphere (=1013mbar) sauses a decrease of the water volume by 5.3x10-5 of the original volume

h=150m

ΔP= gh+1patm
=10^{3}x10x150+105
=16x105 N/m^2

compress=(16-1)x5.3x10-5
=0.0008

-150=1.0008x103 kg/m^3

ΔV=0.0008x150m^3
=0.12m^3

A
energy generating zone
H
B
main functional zone
C
underground refresh zone
gas O
water
bettery floor
pax propellor
balance water tank
balance water tank
main tank 0~24m^3
vertical driven force
main tank
vertical driven force
12 horizontal balance water tank

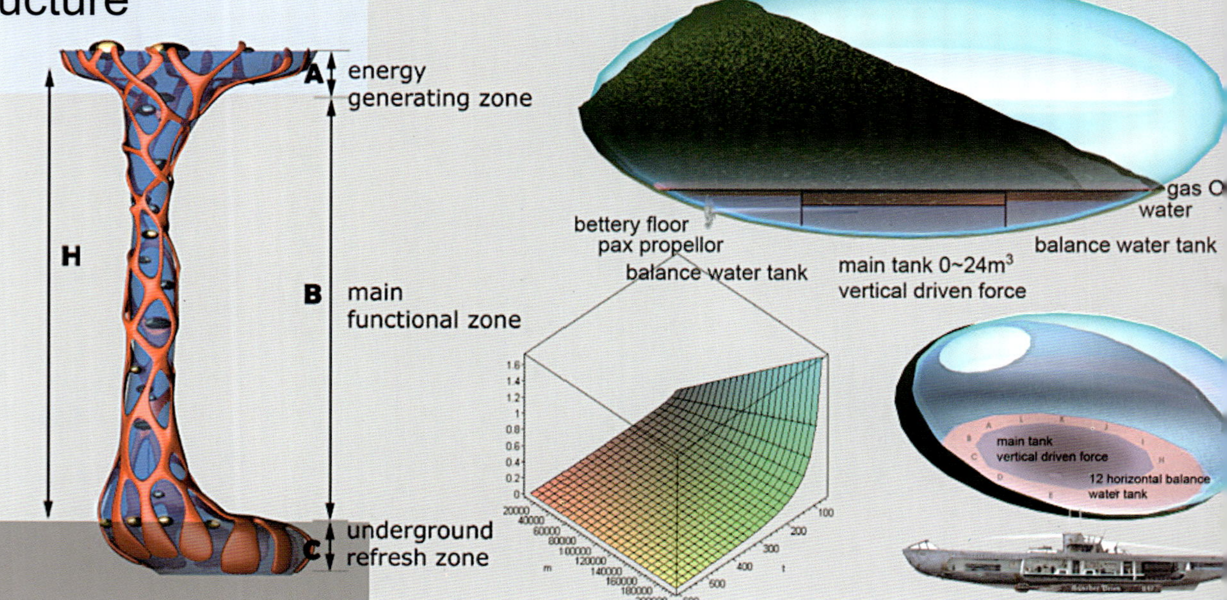

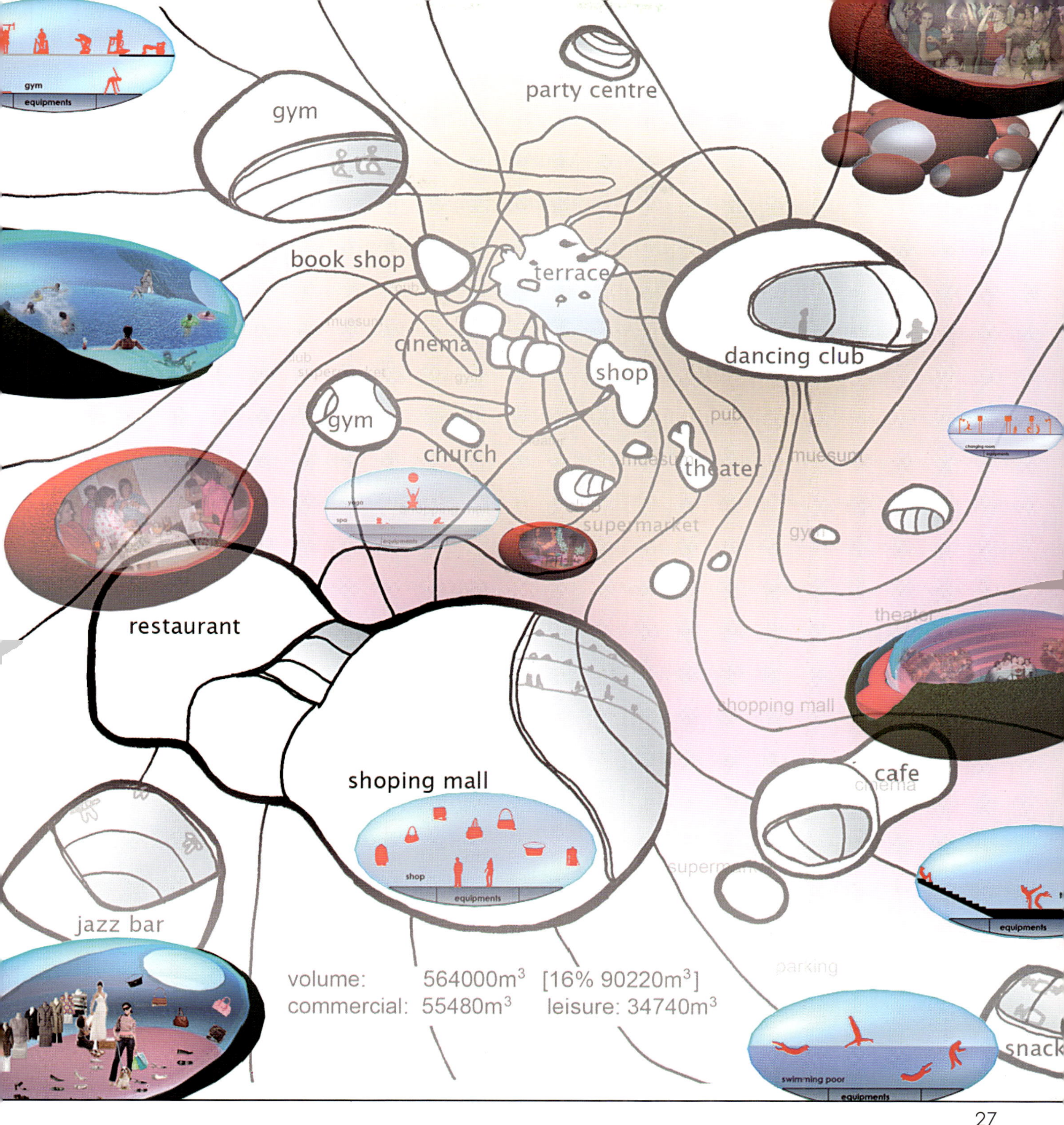
gym
equipments
party centre
gym
book shop
terrace
dancing club
cinema
shop
gym
church
pub
muesum
theater
muesum
yoga
spa
equipments
super market
gym
restaurant
theater
shopping mall
shoping mall
shop
equipments
cafe
cinema
jazz bar
volume: 564000m³ [16% 90220m³]
commercial: 55480m³ leisure: 34740m³
parking
swimming poor
equipments
equipments
snack

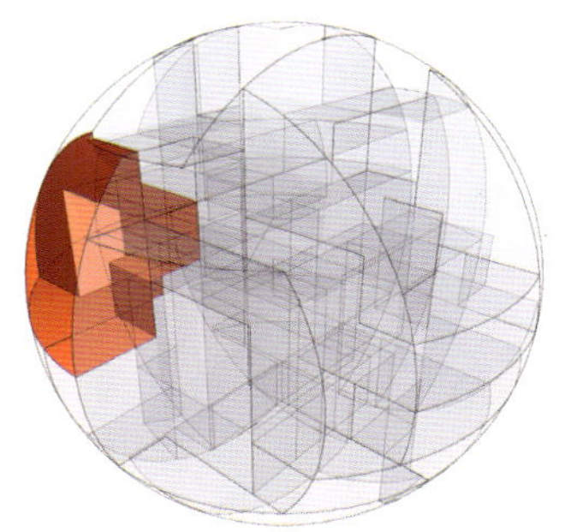

Zone 2 - Urban Nomads
都市流浪者
Christina Rangel

我们像都市的流浪者一样生活——我们实际上在工作中生活，我们当中的一些人甚至在工作场所吃饭、睡觉和休息，因此表明功能和场所是可以互换的。这就是第二区的主要设计思想。通过连续的屋顶、墙体和楼面，工作、生活、睡觉连续性的循环被贯彻在办公空间的形式和活动的居住单元当中。线性的办公空间挑战了垂直叠加的办公层。开放的平面允许公众与工作环境相互作用引导空间使得空间更具可视性和可达性，因此产生出一种类似街道的气氛。没有空间的分级，因此空间能够根据需求而改变。居住单元容纳了工作和居住的功能并能够根据个人需要分隔或交叠。单元沿着外立面的运动对于外界显示出不断变化的外观。第二区的位置靠近球体的顶部，面向东南。这一位置可以利用太阳能进行可持续设计。表皮可通过收集太阳热量产生能量以达到零能量的球体的目标。连续楼面提供了结构稳定性，因此只需要最少的支撑并减少对下部的垂直荷载。这也允许自由流动的开放工作空间。

We live like Urban Nomads - we work from home, we practically live at work, some of us even eat, sleep and rest at work. Hence it is seen that function and place are interchangeable. This is the main design idea for Zone 2 the continuity of the work/live/sleep cycle is carried through into the form of the office spaces and mobile residential units by the uninterruption of roof, walls and floor.

The “linearity” of the office spaces challenges the more common “Vertical Stacks” of office levels. The open plan allows the public to interact with the work environments while navigating the space making the spaces more visible and accessible thereby creating a “street-like” atmosphere. There is no spatial hierarchy so spaces can be modified based on need. The residential unit accommodates working and living functions that can be seperated or overlapped depending on the needs of the individual. The movement of the units along the facade gives the exterior an appearance of constant change. Zone 2 is located towards the top of the sphere and faces south-east. This location allows for sustainable design that takes advantage of the heat from the sun. The skin is designed to generate energy by Concentrating Solar Thermal Power thus contributing to the goal of a ‘Zero-Energy’ Sphere. The continuous floor plate allows for Structural Stability thus needing minimal support and reducing the vertical load on the plots below. This also allows for free flowing, open, work spaces.

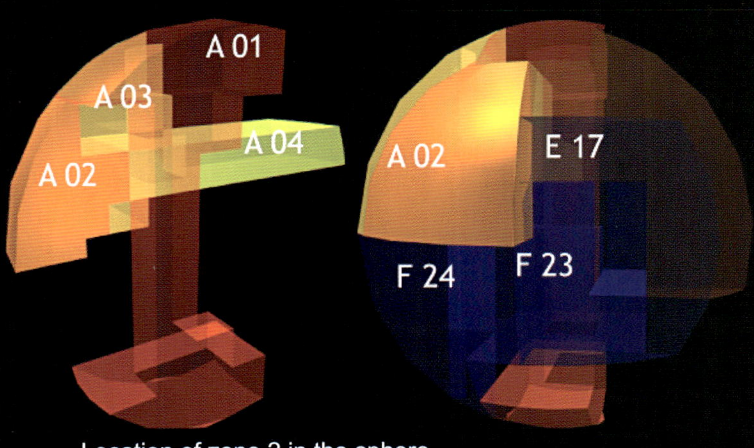

Location of zone 2 in the sphere

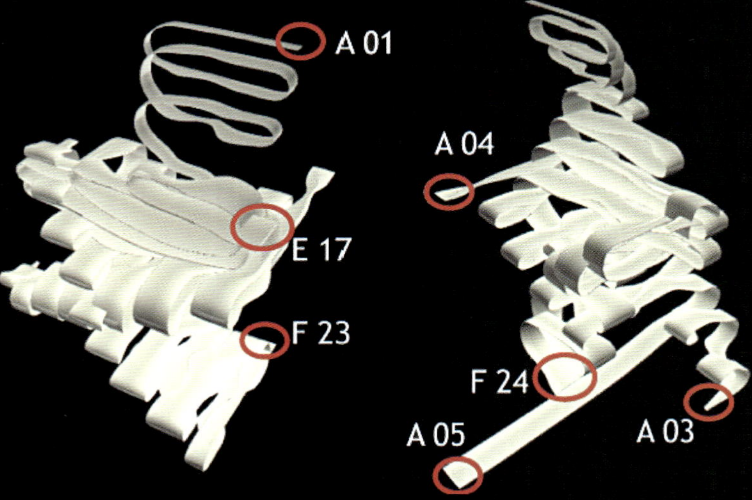

Neighbouring connections

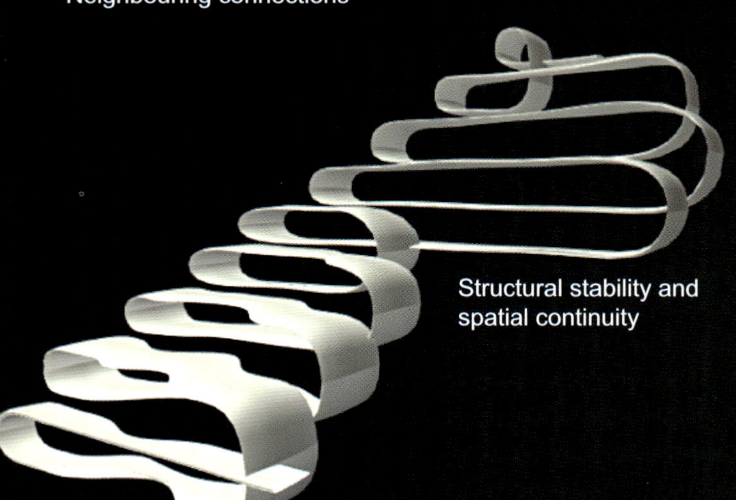
Structural stability and spatial continuity

Nomadic lifestyles

Urbanity

Vertical Stacks of Office Spaces

Model testing the structural integrity of form and material

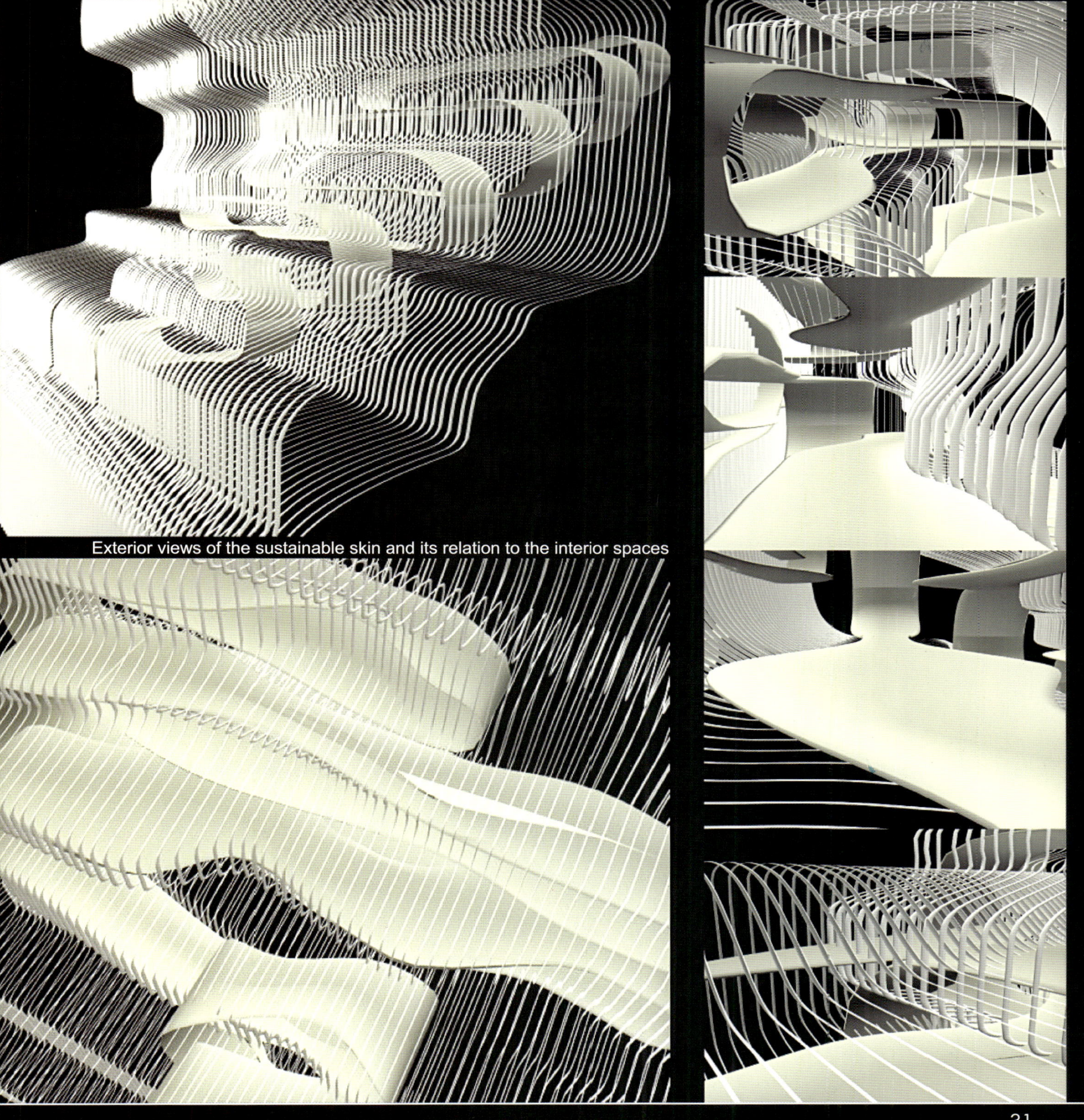

Exterior views of the sustainable skin and its relation to the interior spaces

Interior views of the spaces

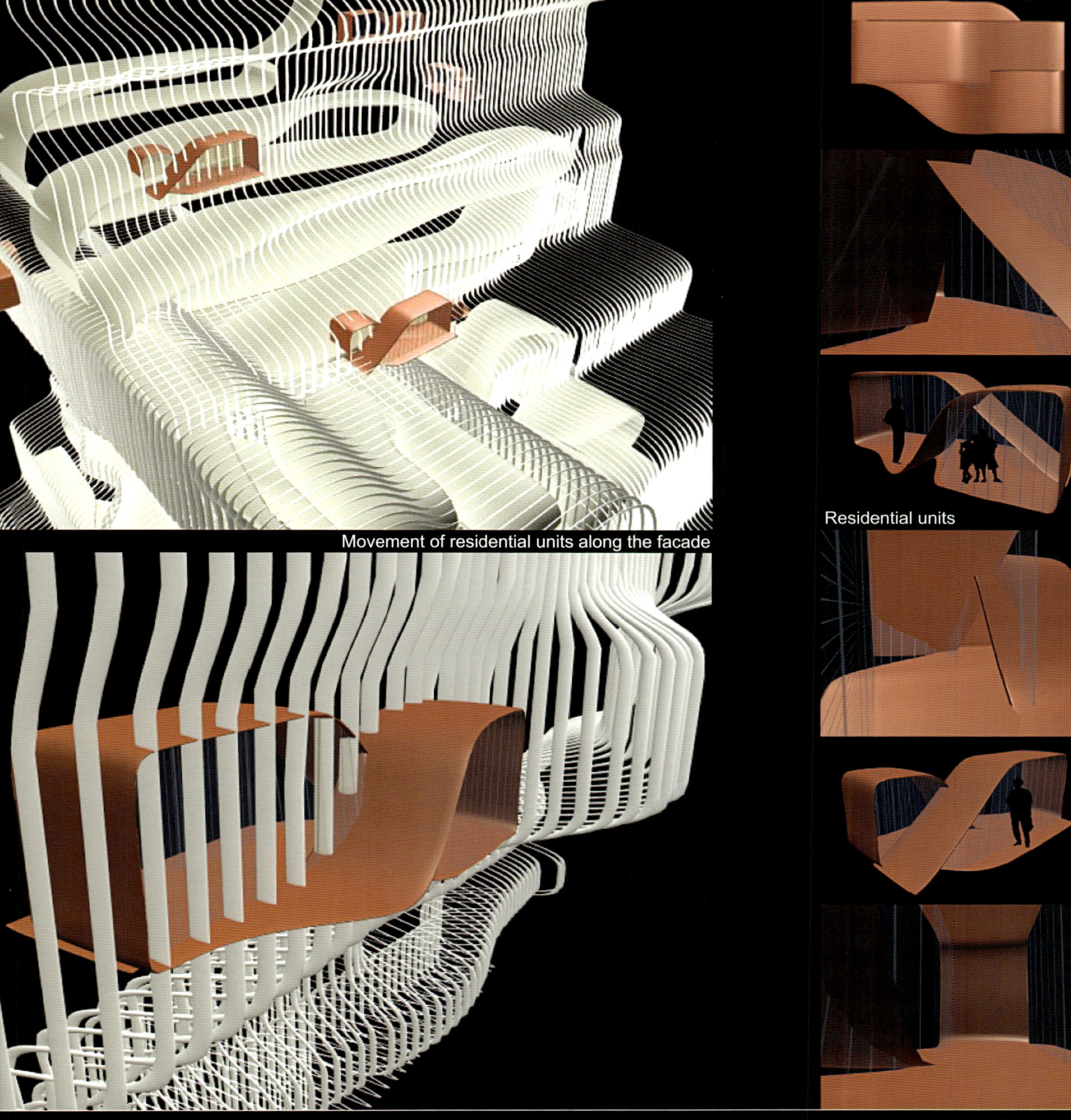
Residential units
Movement of residential units along the facade

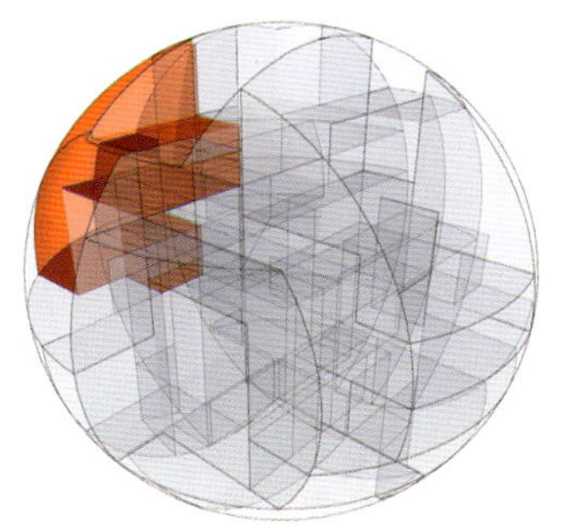

Zone 3 - The Wireframe Spheres

线框球体

Marije van der Laag

在北京751球体的顶部，一个钢铁的线框建筑通过两个85米高的交通塔支撑起来。这两个塔连接着相邻地块。线框的表面供仓体用来从A移动到B。它面向西南方，包含着住宅和旅馆。仓体通过机械联系来悬挂在表面。通过旋转和线性运动，仓体不断调整，以适应气候和日照条件。它们将整日随着太阳的运动而变化。通过使用两层立面结构之间的球形轴承装置，仓体内部的旋转成为可能。通过此装置的使用，内层将始终保持处于准确位置。整个立面都是玻璃窗，在玻璃内部注入了微型电池，以收集光照，产生能量。这些电池彼此连接，众所周知，该系统为Sphelar，它将为高阶照片提供高效率的电子转换，这些电池可用于曲面玻璃中。

In the top of the Beijing 751 sphere, a steel wire frame construction is supported by two 85 meter high transportation towers. The towers are connected to the neighbor plots. The surface of the wire frame is used by capsules to move from A to B. It is facing a south west orientation. The capsules contain dwelling and hotels. Through a mechanical connection, the capsules are hanging on the surface. The capsules are constantly adjusting to the climate and sunlight condition, by rotation and linear movement. Throughout the day, they will turn with the movement of the sun. The rotation within the capsules is possible by the use of a ball-bearing mechanism between two layers of façade construction. By the use of this mechanism, the inside layer will always stay correctly leveled. Glass windows are filling the façade. Inside the glass, miniscule cells are injected to collect sunlight and regenerate energy. The cells are wired to each other. This system is known as Sphelar. It is providing a high level of photo electric conversion efficiency. These cells can be used in the curved shape of the glass.

outside skin starts to rotate

but internal skin keeps horizontal position

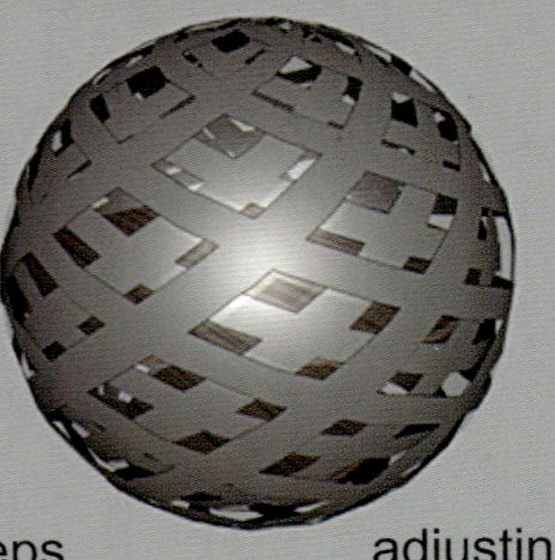

adjusting structure

capsule
d=6m

2 layers
living

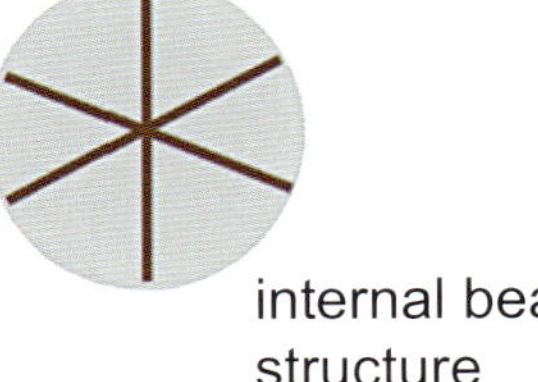

internal beam structure

beams
& floors

outside structure

roof & bottom view

interior: bamboo
exterior: steel and glass with solar cells

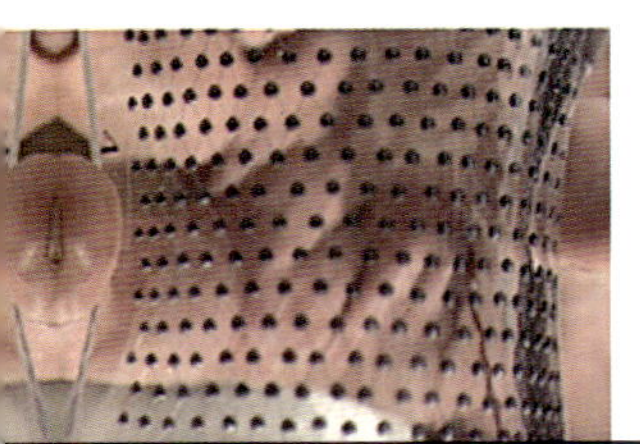

wired spheral flexible module

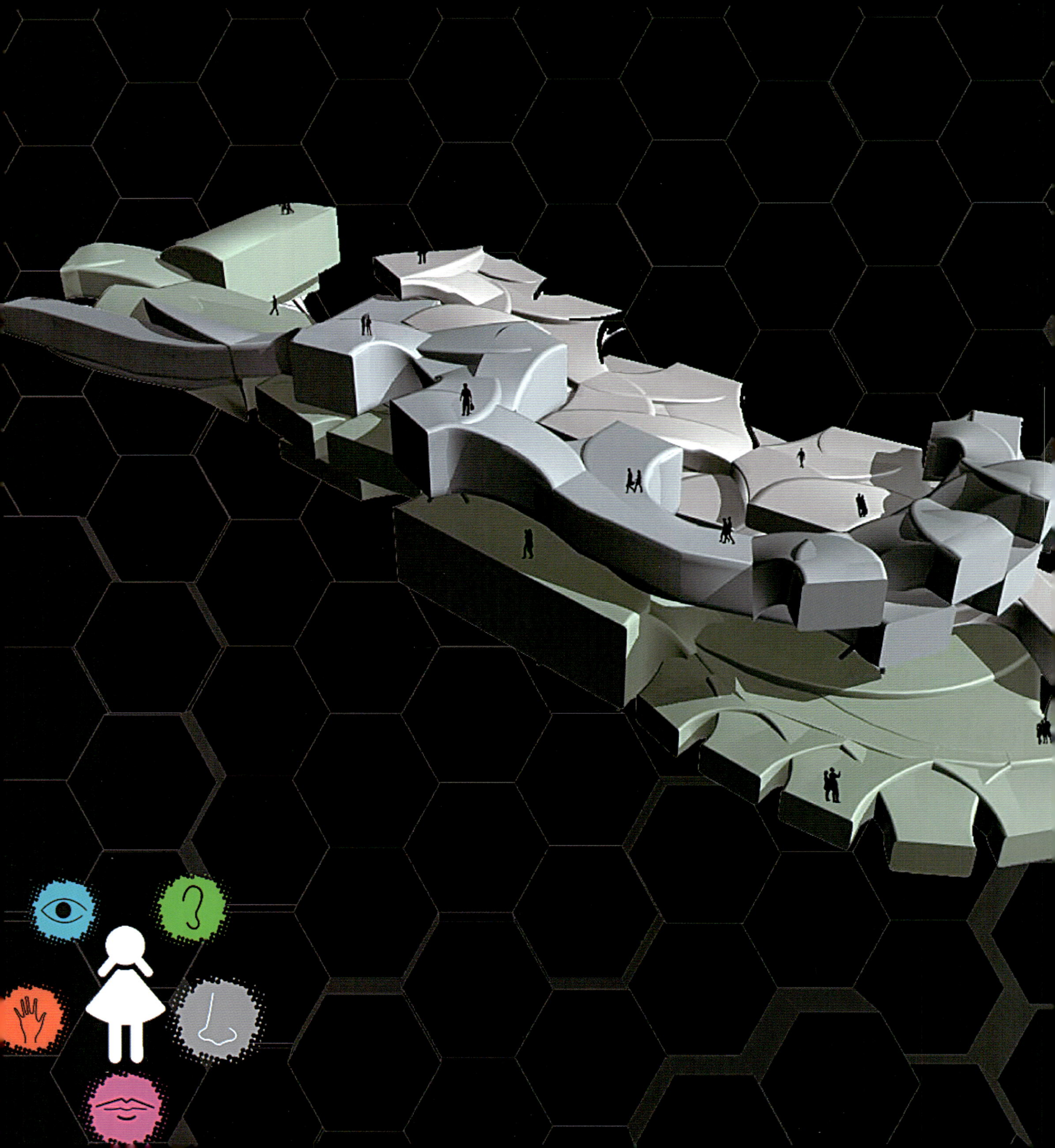

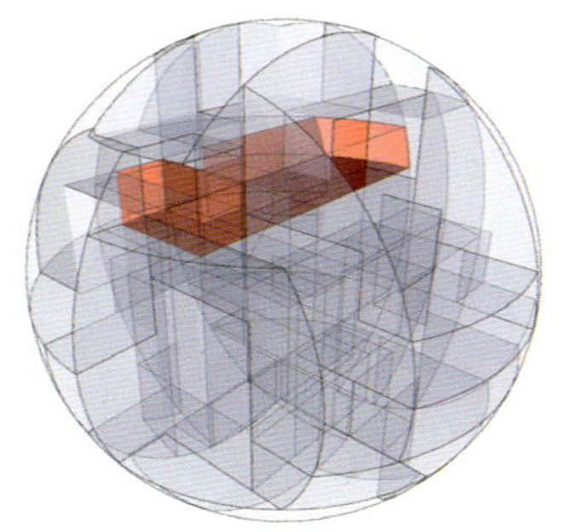

Zone 4 - High Five
强五感
Nora Schuler

该项目名为“强五感”，位于球体的第4区。由于该区是水平结构，所以建筑也在水平方向上被拉伸，安置了一个适合751艺术区概念的中国美术研究综合体。它包括四种功能：一个博物馆、一个小学、一个研究所和学生宿舍，每项功能在建筑内部都拥有一条自己的通道。

建筑造型源于建筑内部人流。各功能未连成一排，而是移向一边和类似入口的大路，不是那种会带来很大人员流动的矩形入口。这样，外壳就根据这种流动，用简单的圆边矩形来建造。房间结构的不同可能性可总结为一系列模块。根据功能需要来组装模块便可产生最终造型。人在房间外部和内部都可以步行。

“强五感”的名称来自室内的概念，这个概念也形成了外部与内部的反差。正如在日常世界中，外部是基于更加视觉化的成分。而内部则应当通过将他们彼此分离，单独呈现，来处理好所有的感觉。对于一项功能的每个房间来说，四到五项感觉汇集起来，以使参观者进入之前，不是用真实的空间结构，而是用感觉体验属于哪种功能的解释，来接收到建筑的地图。所以，例如通过辨认一个房间的嗅觉、味觉和触觉，参观者就能找到该房间。如果辨认太困难，则参观者会返回室外。

The name of the project is “high five“, and it is located in zone number four of the sphere. Due to the horizontal configuration of the zone the building is stretched in the same direction, inhabiting an academy complex of Chinese fine arts that will fit into the 751 art district concept. It consists of four functions: a museum, an elementary school, an academy, and student houses and each function with an own path inside the building.

The shape of the building derives from people flows in the inside. The functions are not chained in a row but moved to the side and highway resembling entrances instead of rectangular entrances ensure a high intense flow and movement of the people. The outer shell is then built by following these flows with a simple round edged rectangle. The different possibilities of room configuration can be summarized in a catalogue of modules. Assembling the modules according to the function needs will lead to the final shape. It is possible to walk both in the inside and on the outside.

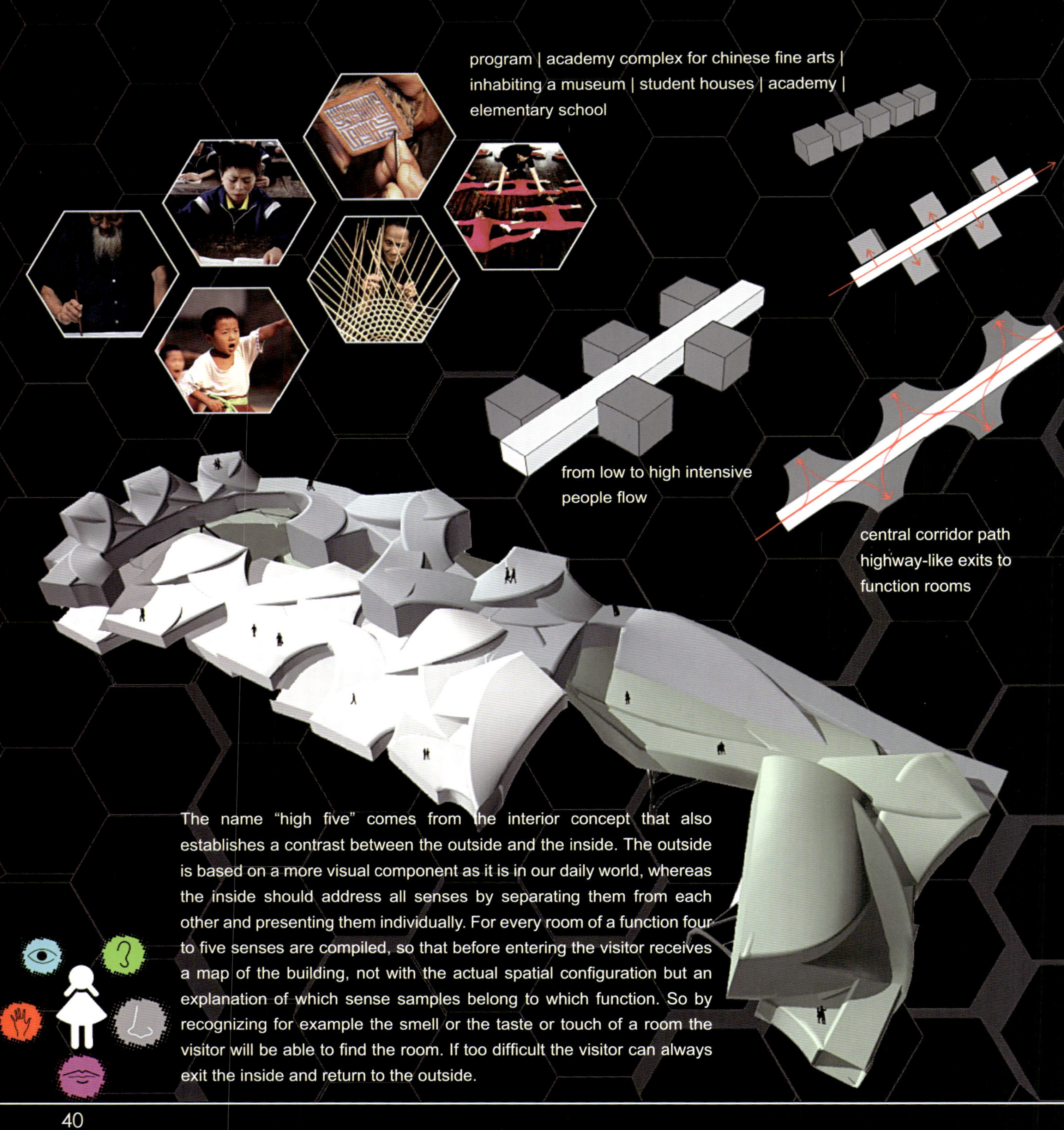

The name "high five" comes from the interior concept that also establishes a contrast between the outside and the inside. The outside is based on a more visual component as it is in our daily world, whereas the inside should address all senses by separating them from each other and presenting them individually. For every room of a function four to five senses are compiled, so that before entering the visitor receives a map of the building, not with the actual spatial configuration but an explanation of which sense samples belong to which function. So by recognizing for example the smell or the taste or touch of a room the visitor will be able to find the room. If too difficult the visitor can always exit the inside and return to the outside.

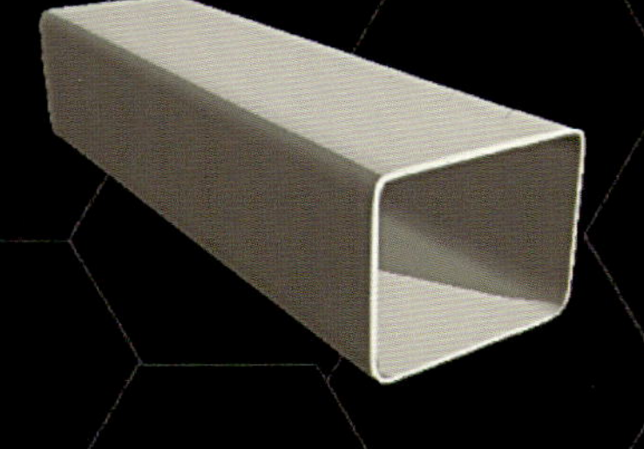

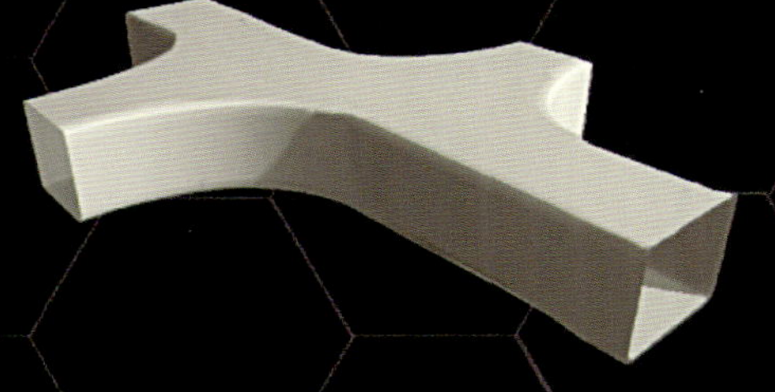

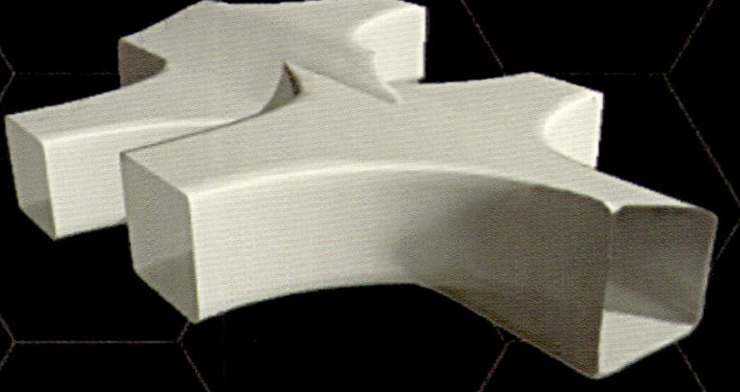

corridor |
senses guidance system

with functions on the side
on corridor level

functions on second level
exits to the outside

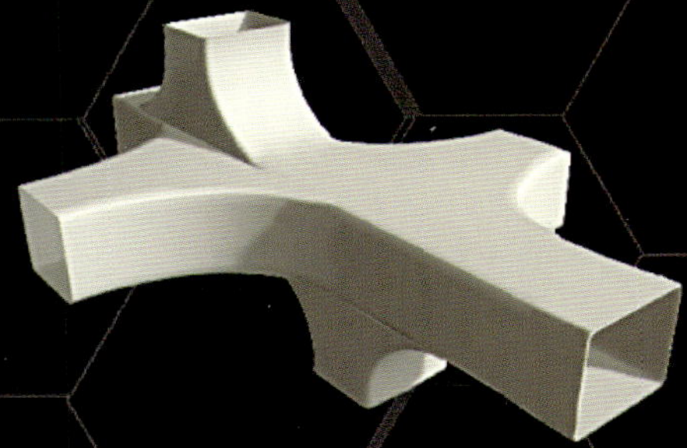

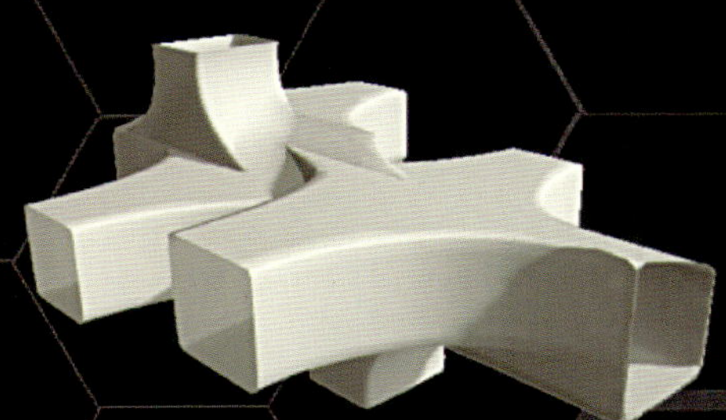

prototype |
senses guide

functions on top and bottom

combination of all modules

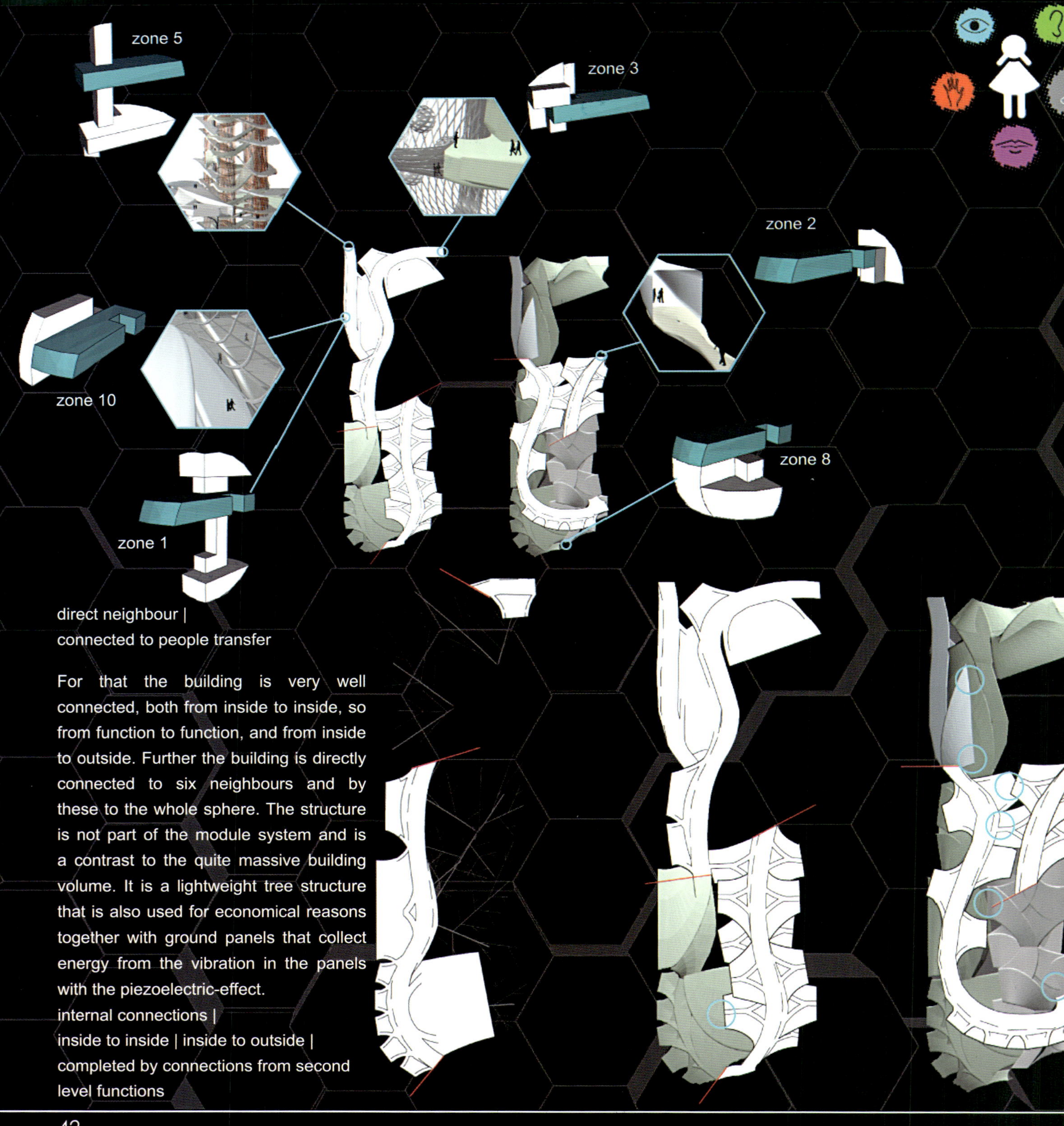

direct neighbour |
connected to people transfer

For that the building is very well connected, both from inside to inside, so from function to function, and from inside to outside. Further the building is directly connected to six neighbours and by these to the whole sphere. The structure is not part of the module system and is a contrast to the quite massive building volume. It is a lightweight tree structure that is also used for economical reasons together with ground panels that collect energy from the vibration in the panels with the piezoelectric-effect.

internal connections |
inside to inside | inside to outside |
completed by connections from second level functions

structure|
aluminium filled with glassfibre
concrete
lightweight tree structure
three parts carry vertical loads of 121000kN | starting diameter of trees is 2m
no horizontal forces are considered
modules |
glassfibre plastic
embedded stencils for sensing panels
energy gain |
piezoelectric effect
from vibration in
surface panels
lighting |
hybrid lighting
with glassfibres
collected sunlight
and electricity
gained from
piezoelectric effect

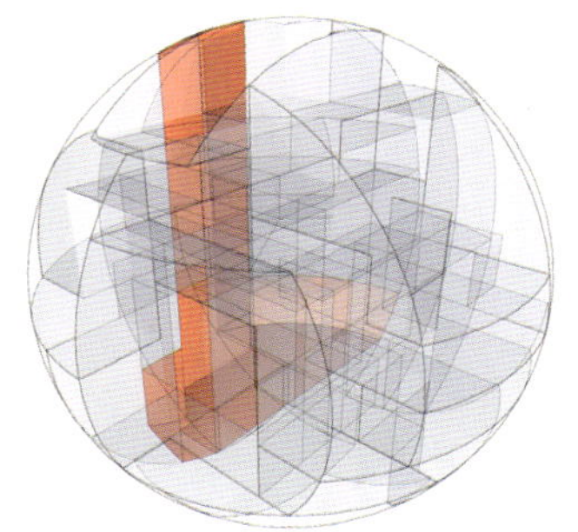

Zone 5 - Orientation-Transport-Support

方位——交通——支持

Miroslava Tumova

第5区位于球体的中心部分。垂直瘦长的形状直抵顶端和底端，其水平部分在地面以上10m。该地块有许多相邻地块。球体的这个部分应当提供定位服务，并通过人员、能量、车辆、也包括构筑物的运输来支持所有相邻地块。

建筑方案占据了整个地块，并保持内部的巨大开放空间，该开放空间应作为一个定位核。人们来到球体，将他们的汽车放在地块水平部分的自动停车库中。然后参观者或者居民继续乘快速电梯上到球体中。在巨大的交通塔的每一点，他们都能选择四种不同的方式来继续行程。他们可以在公园途中放松休息，或者干脆立刻进入球体的其他部分。

整个建筑被设计为四个中空钢柱支撑下的空间混凝土梁结构，这个部分提供了球体主要的结构支撑。

Zone 5 is located in a central part of sphere. Thin vertical shape is touching top and bottom and horizontal part is 10m above ground. Plot has a lot of neighbours. This part of sphere should bring orientation and support all neighbours by transportation of people, energy, cars and also structure.

Proposed part of the building fill all provided shape and keep big open space inside. This open space should serve as an orientation core. People come to the sphere and let their cars in an automatic car machine in the horizontal part of plot. Then visitor or habitants continue by fast elevators up to the sphere. On every point of big transportation tower they can choose four different type of way to continue. They can relax on a park way or just hurry up into the other part of sphere.

All structure is designed as a spatial concrete beam with four hollow steel columns. This part of sphere serves as a main structure support.

main orientation core

concept

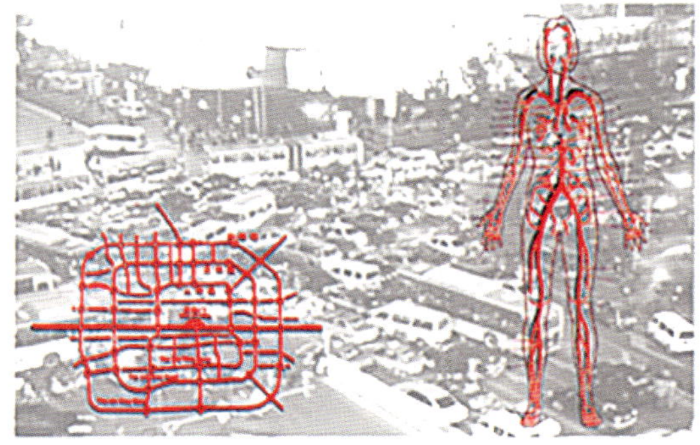

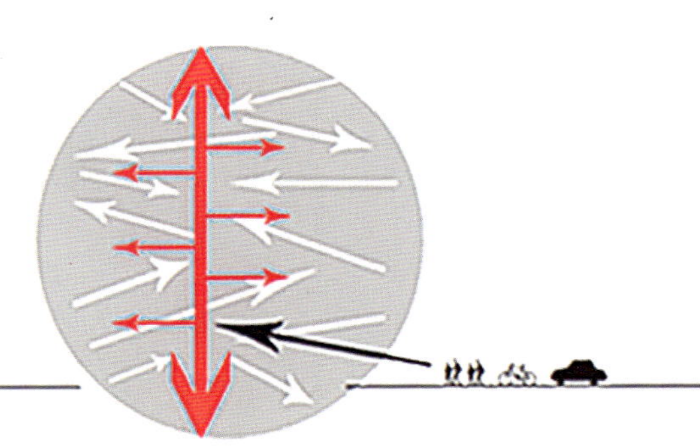

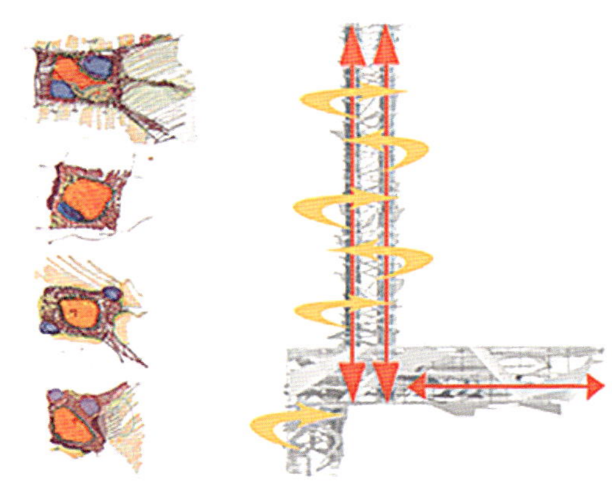

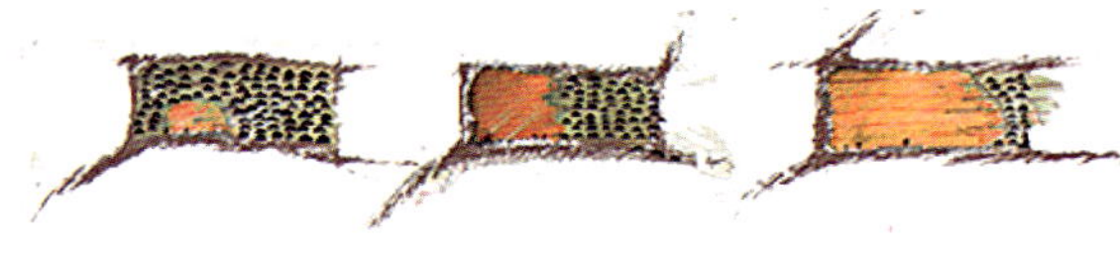

inputs:

position of plot in the
sphere shape of plot
parallel to human body
parallel to Beijing
functions

outputs:

orientation
transport of energy
thorougfare of sphere
support neighbours
public space
storage of car

functions:

vertical transport>tower
parking > 90000m³ >
horizontal part
commercial > 15000m³
offices > 4000m³ >
ways,ramps
leisure > 4000m³ >
ways,ramps

transport:

fast movement > main
public space
four elavators
slow movement >
ramps, ways

car machine:

multiplicate pater noster
let your car park alone!
don´t drive
9 parking floors >
2000 cars
parametr > car units > box

structure:

irregular concrete shell
on the borders
four steell tubes
overlaping concrete
ramps

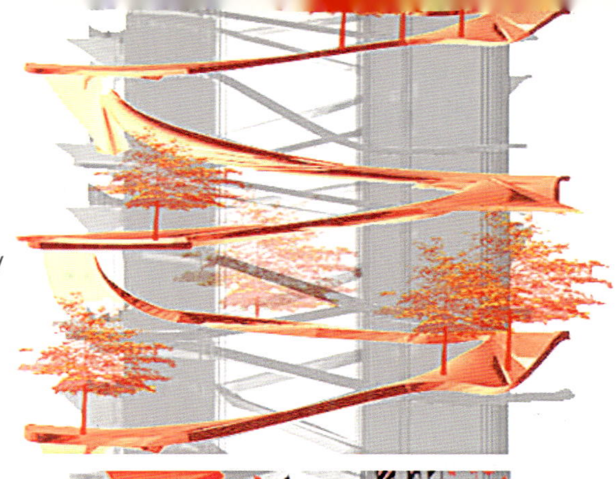

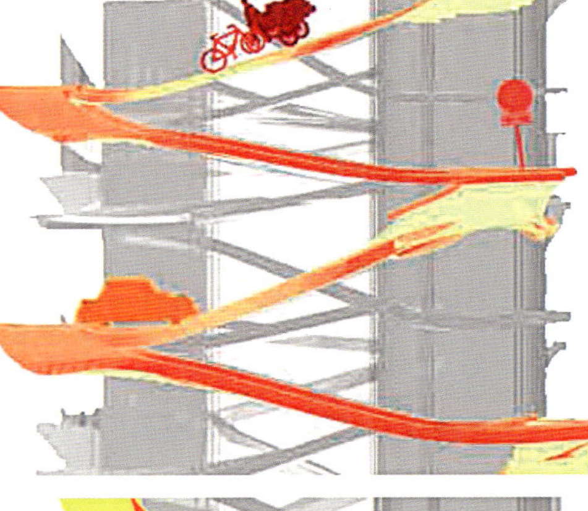

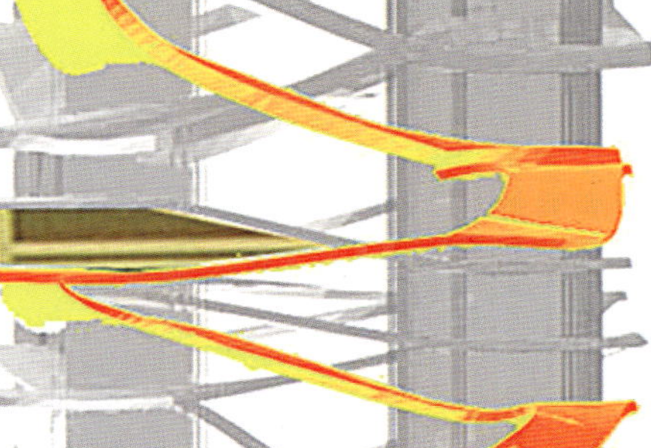

main public space

main connections

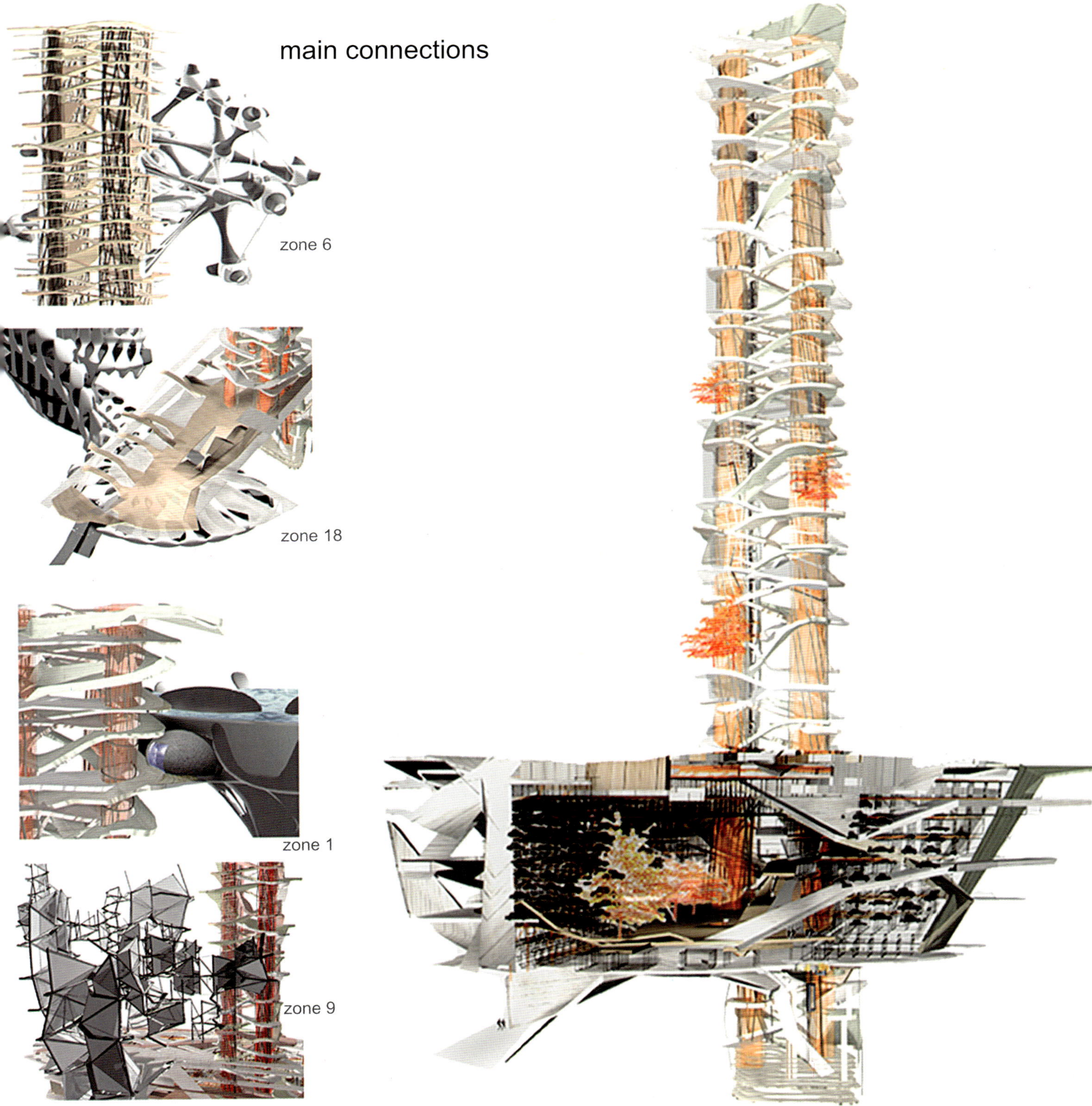

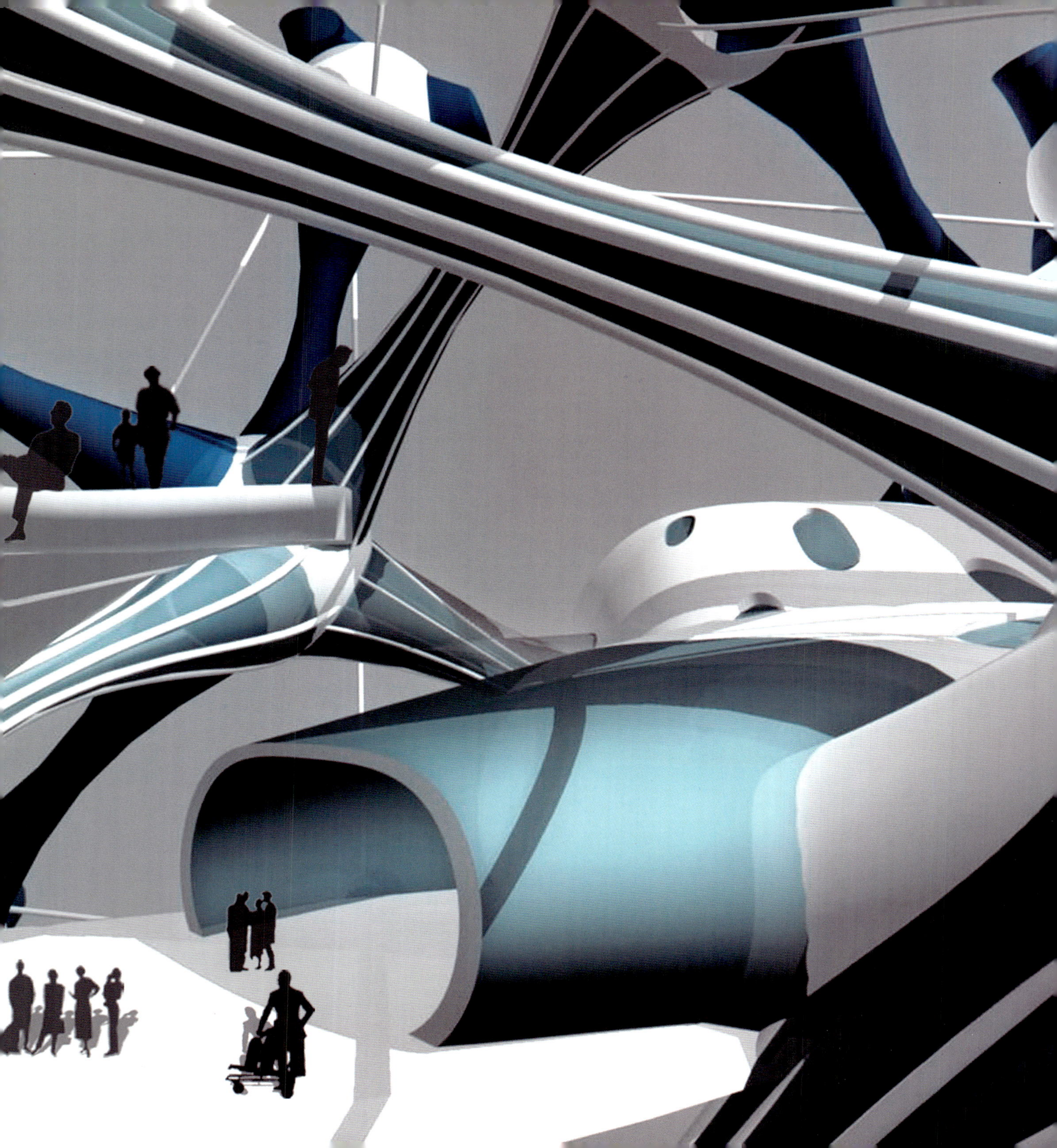

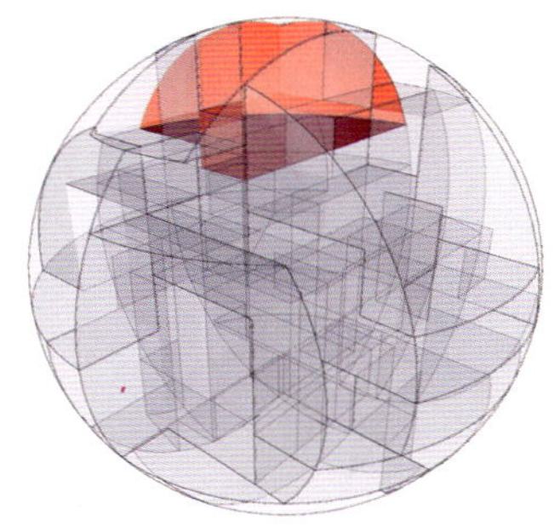

Zone 6 - Hotel+Congress hall
旅馆+会议大厅
Roman Krajger

该旅馆方案的概念基于群集行为方法，这意味着该旅馆中的房间单元是可移动的。天气状况或使用者可以自己决定的几个参数引导着房间的运动。这影响着整个系统：找寻最佳位置。

目标在于主要与第7地块连接，它也是主要的结构支撑，并允许自由通行。所谓的连接器，包括门厅、会议厅，也是主要的入口。

交叉点是固定单元，也是到达各房间单元的起始点。为这些交叉点内的一些目标群设定了一些特定功能。目标群分为：家庭、文化、休闲、运动、娱乐、信息。

每个可移动单元都包括三个房间。该房间有一个固定的内部空间和一个可转动的“管口”。通过转动可形成不同的空间结构：“空间转换器”，整个单元也可转动，双层表皮使之转动，用这种方法也产生能量。这主要是在单元比较空时才出现。

The concept for this hotel proposal is based on the swarm behavior approach. It means that in this hotel the room units are able to move. Several parameters which can be weather conditions or the users can decide personally are directing the movement. This is affecting the whole system:“searching for the best place”.

The object is connecting mainly to plot07 which is also the main structural support and allows the circulation. The so called connector includes the lobby, congress hall and is also the main entrance.
Cross points are the fixed units and are the starting points to reach the room units. There is a specific program for several target groups included in this cross points. The target groups are divided in family, culture, leisure, sport, entertainment, info.

Each movable unit includes three rooms. The room has a fixed inner space and a “nozzle” which is able to rotate. This rotation allows different spatial configurations: “space transformer”.
Also the whole unit is able to rotate. The double layered skin allows the rotation and in this way also produces energy. This mainly can happen when the units are empty.

01 design concept

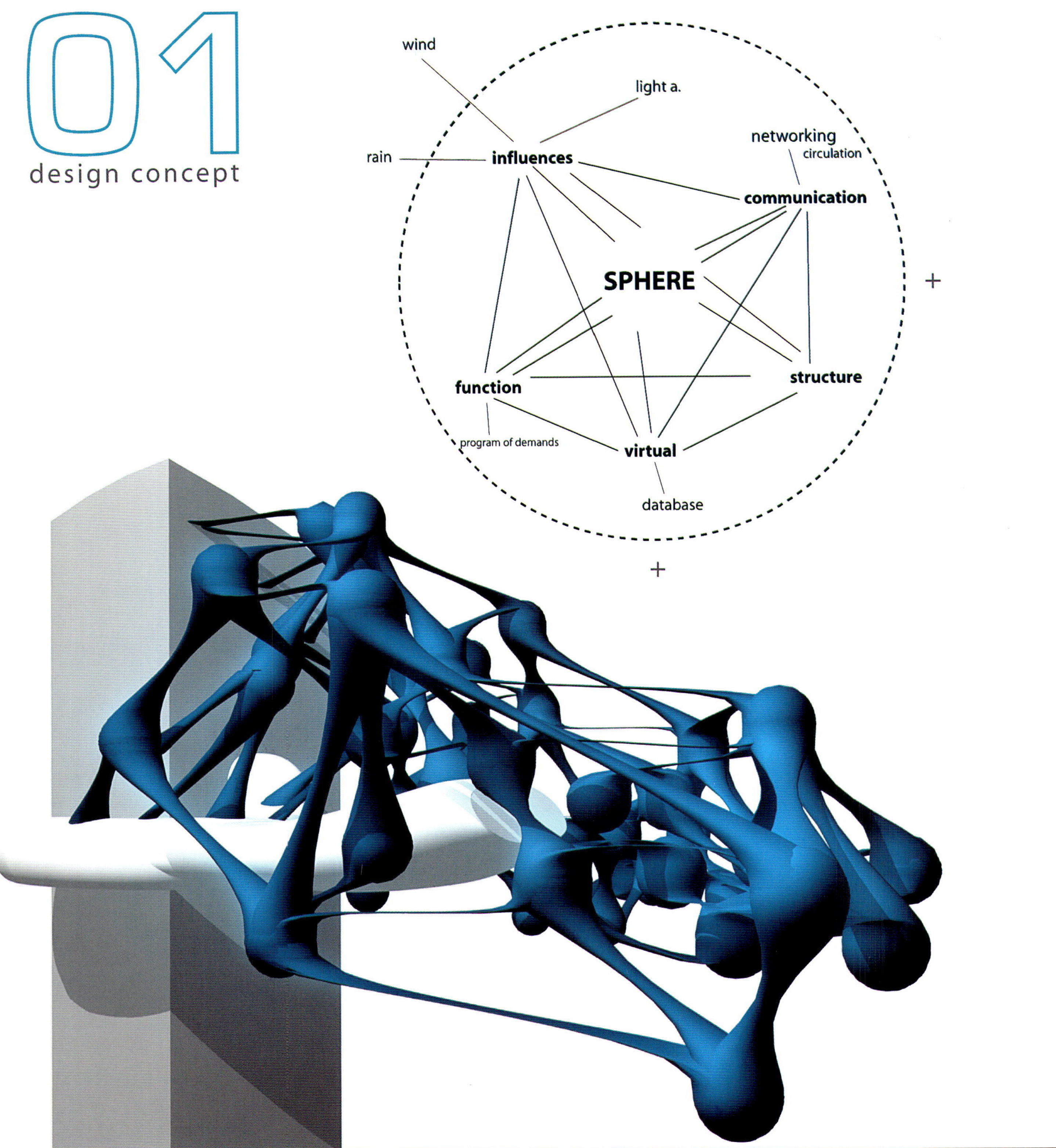

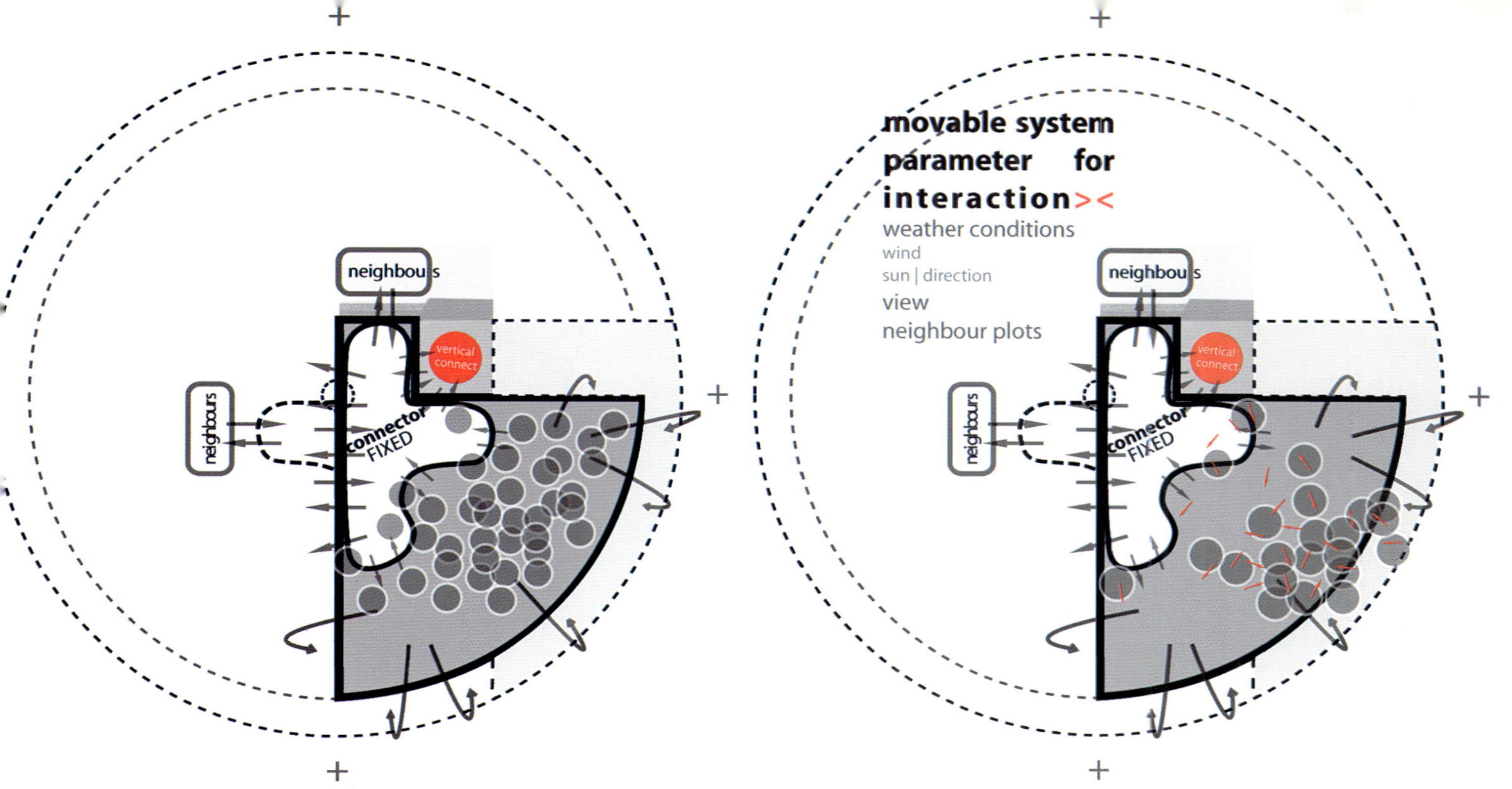

neighbou s

vertical connect

neighbours

connector FIXED

movable system parameter for interaction><

weather conditions

wind

sun | direction

view

neighbour plots

neighbou s

vertical connect

neighbours

connector FIXED

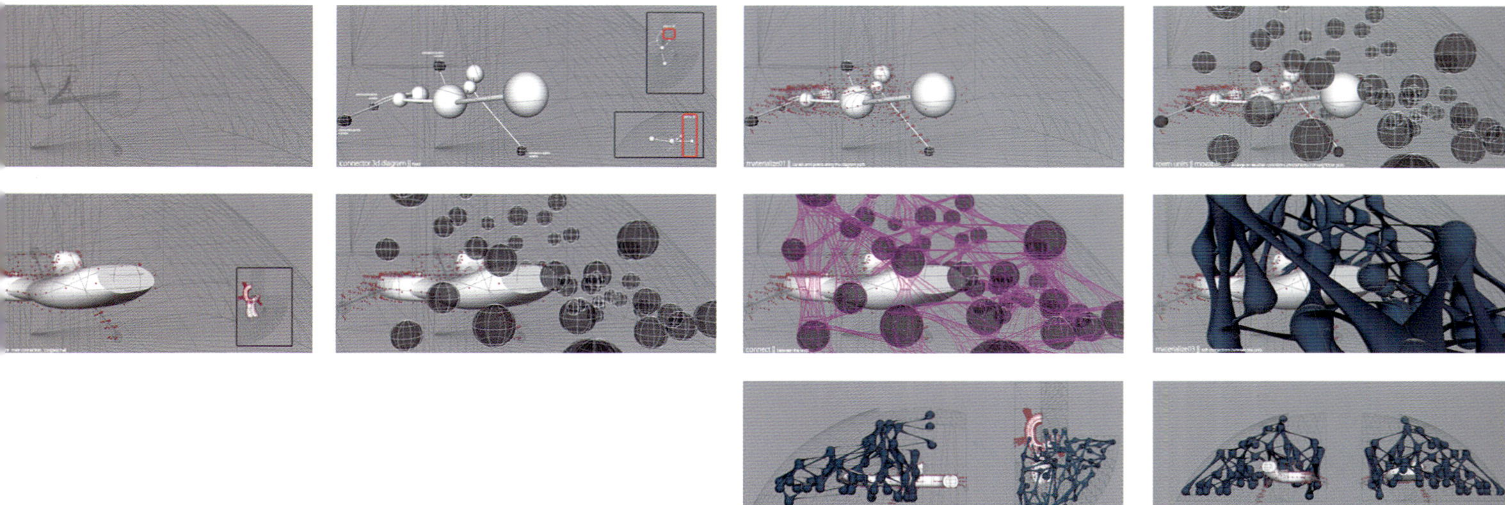

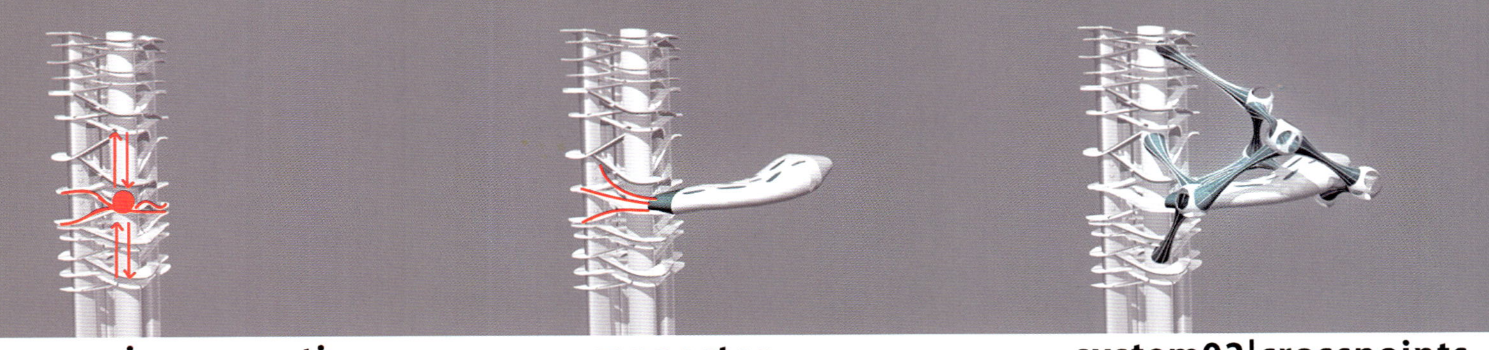

main connection **connector** **system02|crosspoints**

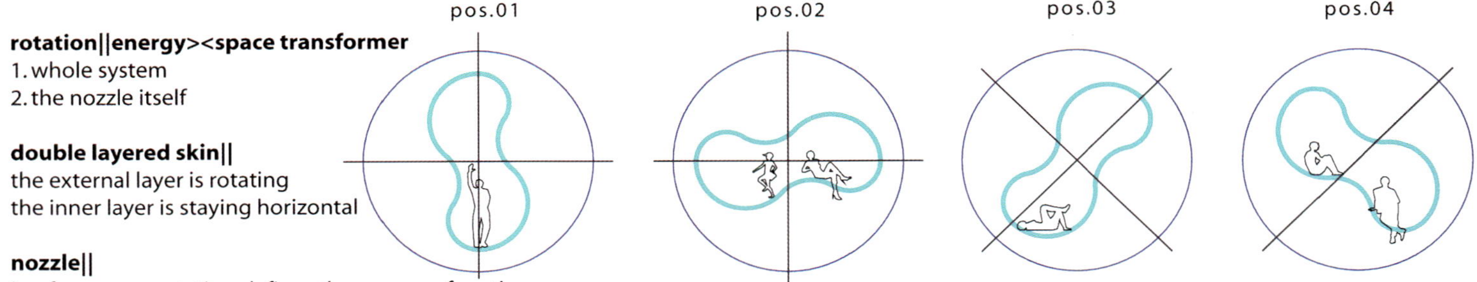

rotation||energy><space transformer
1. whole system
2. the nozzle itself

double layered skin||
the external layer is rotating
the inner layer is staying horizontal

nozzle||
is a free room, rotation defines the space or function...
aerodynamic shape

inside the sphere main functions (bathroom, bedroom, kitchen...)

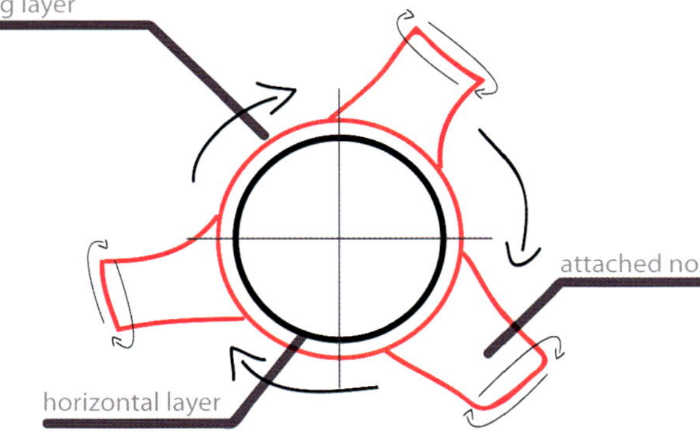

transport system
between the crosspoints escalator
the idea is to have a quick and comfortable transport from crosspoint to crosspoint!
to the room units is planned a magnetic rail
the connection tube to the room units has a diameter about 2.5m.
the transport unit itself is a sphere in wich sould fit at least three people!

02 materialization

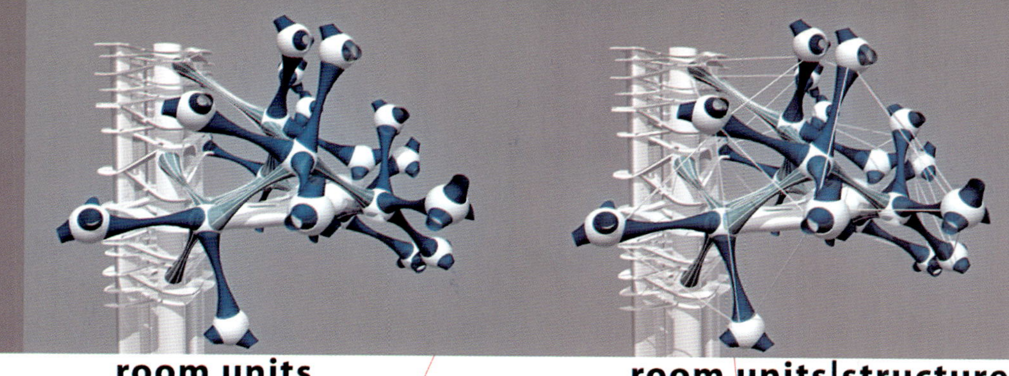

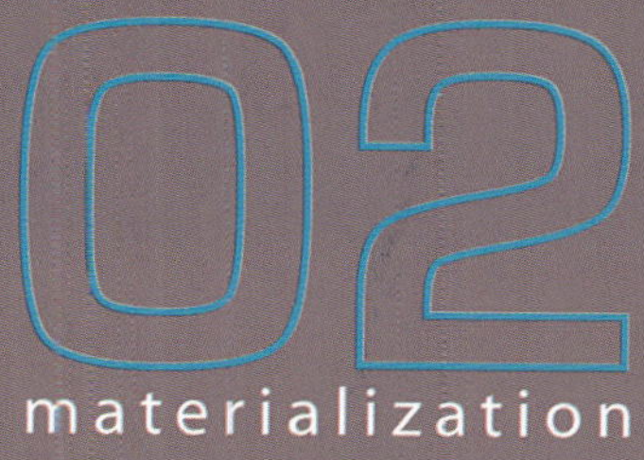

room units

room units|structure

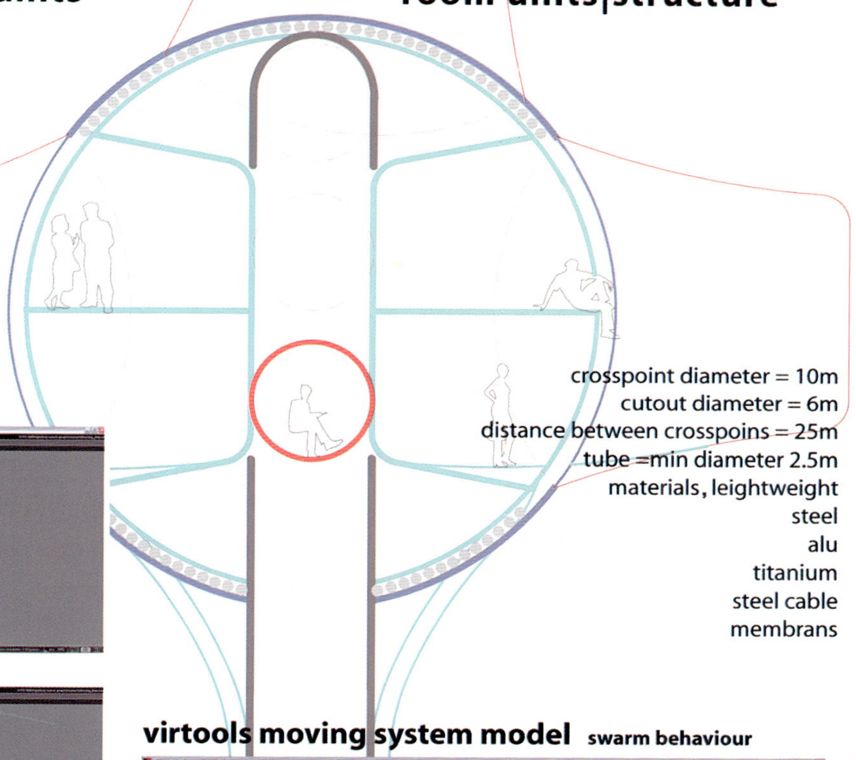

crosspoint diameter = 10m
cutout diameter = 6m
distance between crosspoins = 25m
tube =min diameter 2.5m
materials, leightweight
steel
alu
titanium
steel cable
membrans

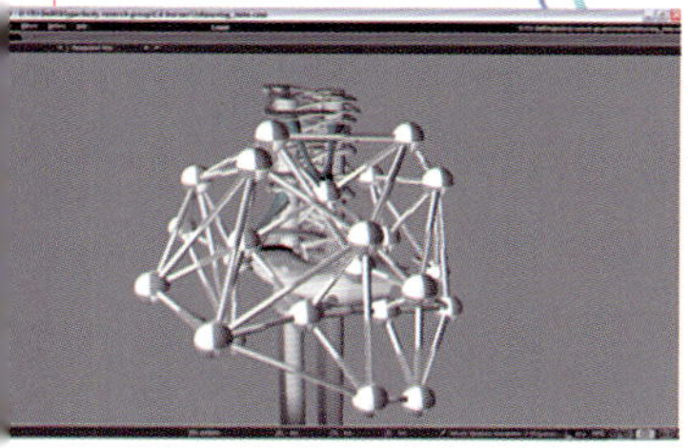

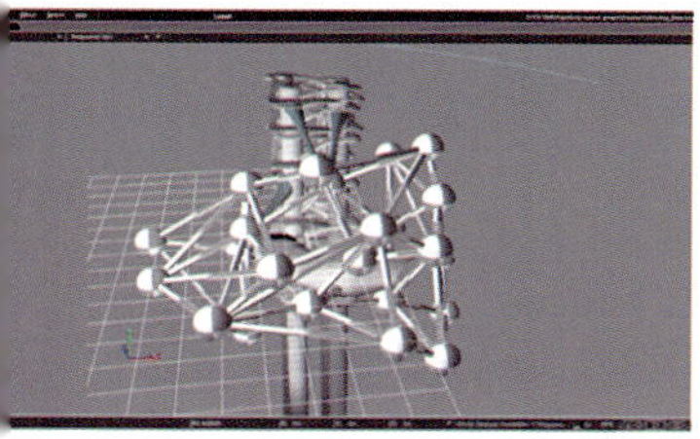

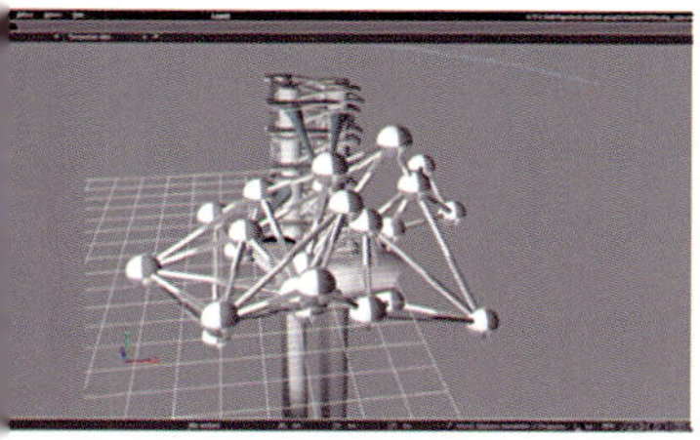

virtools moving system model **swarm behaviour**

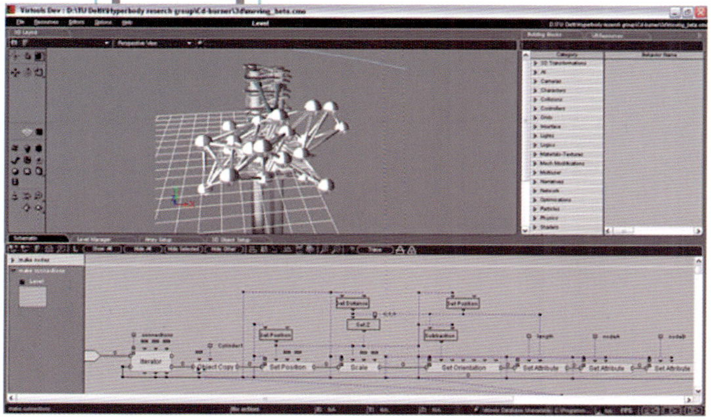

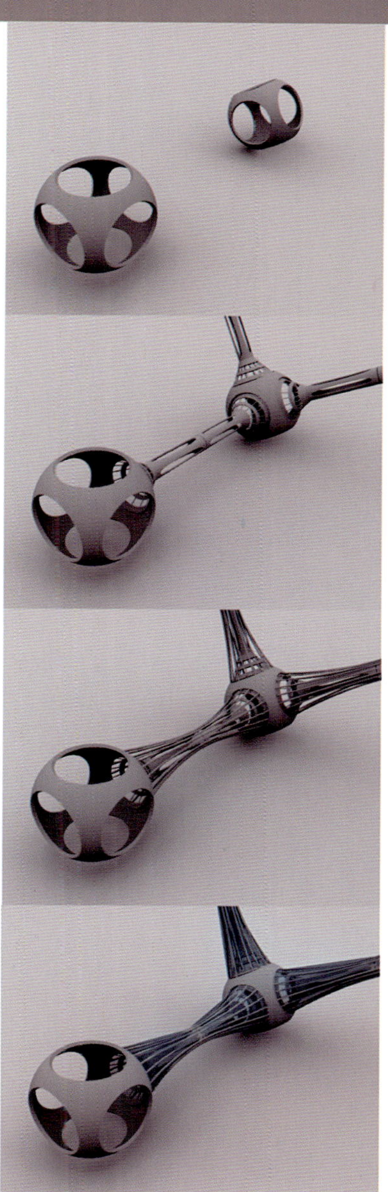

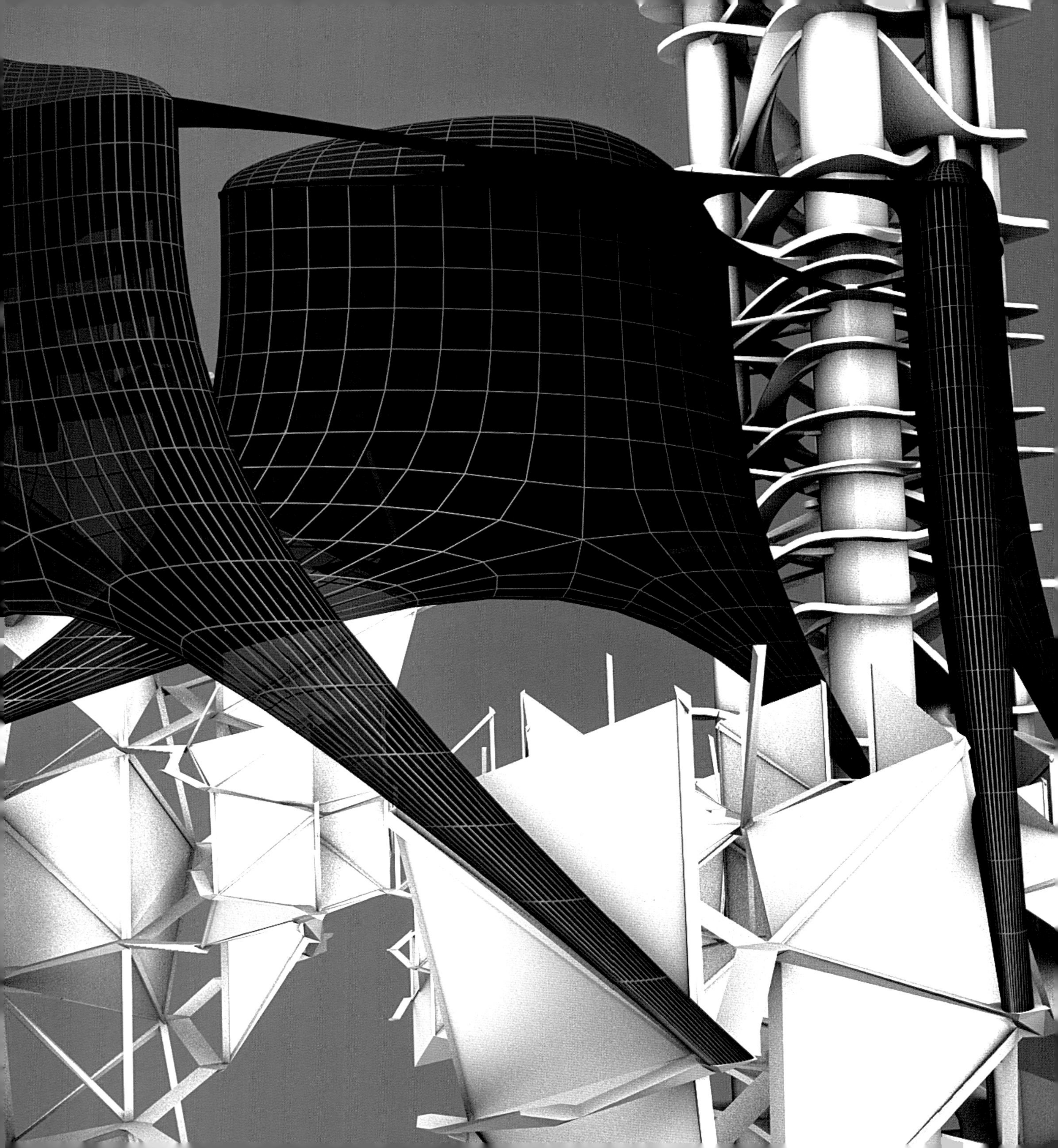

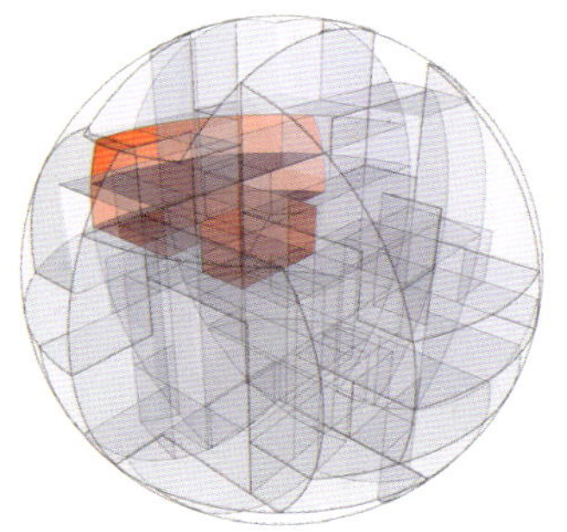

Zone 7 - The Net
网
Christian Mack

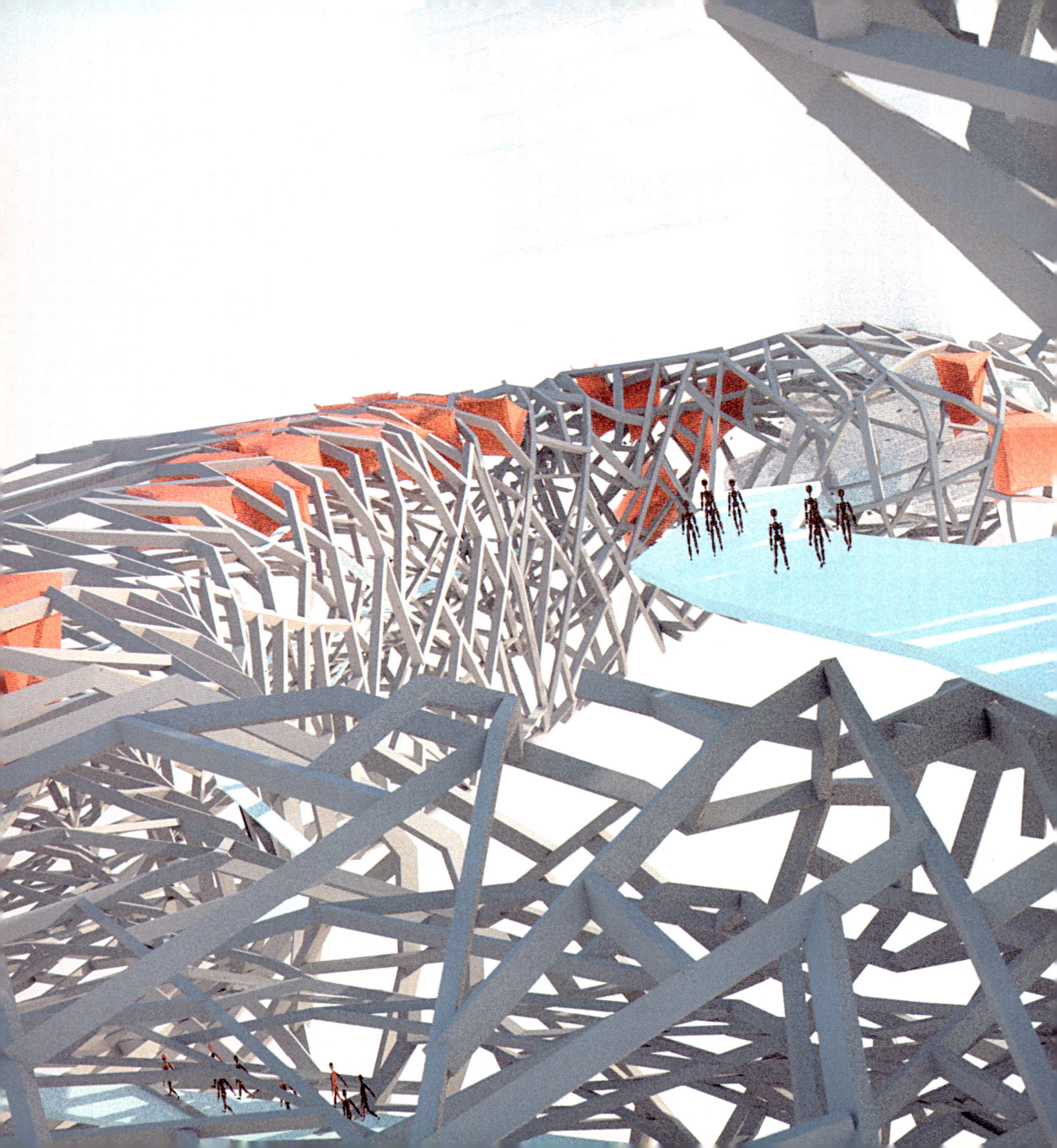

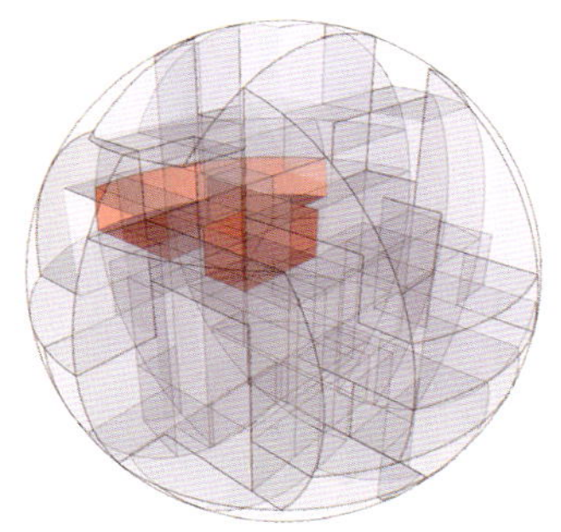

Zone 8 - Camping in 3D
三维野营
Tomas Lapka

北京是每天和每周人口迁入数量最大的城市之一。人口迁入超过迁出十几倍，这导致城市中巨大的人员流动。很多移民是“流浪者”，没有住处。无家可归就成为一个急迫的问题。

主要概念就是要解决上述社会问题：移民，廉价住房。这个想法是要设计居住区的环境，并提供与其他设施的交流。居住区内的组织由人们自己根据局部规则和他们的需要来进行。这种想法来自野营的组织，作为一种临时居住的方式。人们购置自己的帐篷，来到营地，想在哪儿搭帐篷就在哪儿搭，想停留多久就停留多久。

结构设计为一个空间钢框架结构，并转换为三个挤压成的椭圆形或半椭圆形表层。在这些表层之间坐落着居住区，这样整个表面布满了帐篷单元。帐篷具有很好的适应性和可移动性。因此居住是灵活的。在椭圆核中，沿着整个挤压的形状布置着公共空间和步行区。在结构梁之间悬挂着很容易安装的桥，通过这些桥，主要步行通道将各居住区连接起来。

Beijing is one of the cities with the biggest people in-migration per day and week. An in-migration overgrows out-migration more than ten times, which leads to vast people's flow in the city. Many of migrants are “nomads” and do not have any access to living. Thus happens that homelessness is rapid problem.

Main concept is to solve mentioned social issues: migration, cheap access to living. The idea is to design the context of residential area and provide communication with other facilities. Organization within the residential areas is organized by people themselves, by local rules, by their demands. This idea comes out from organization of camping, as a manner of temporary housing. People buy their own tent, come to the camp and build wherever they want, stay as long as they want.

The structure is designed as a steel spatial frame structure which is displaced in three extruded oval or semi-oval layers. In between the layers the residential zone is located, thus can happen that whole surface is filled with tent units. Tents are easily adaptable and removable. Living thus happens flexible. In the oval core, along whole extruded shape there is the public space and pedestrian zone. Main walking line is connected with residential zones with easily installable bridges which are hanged in between the structure beams.

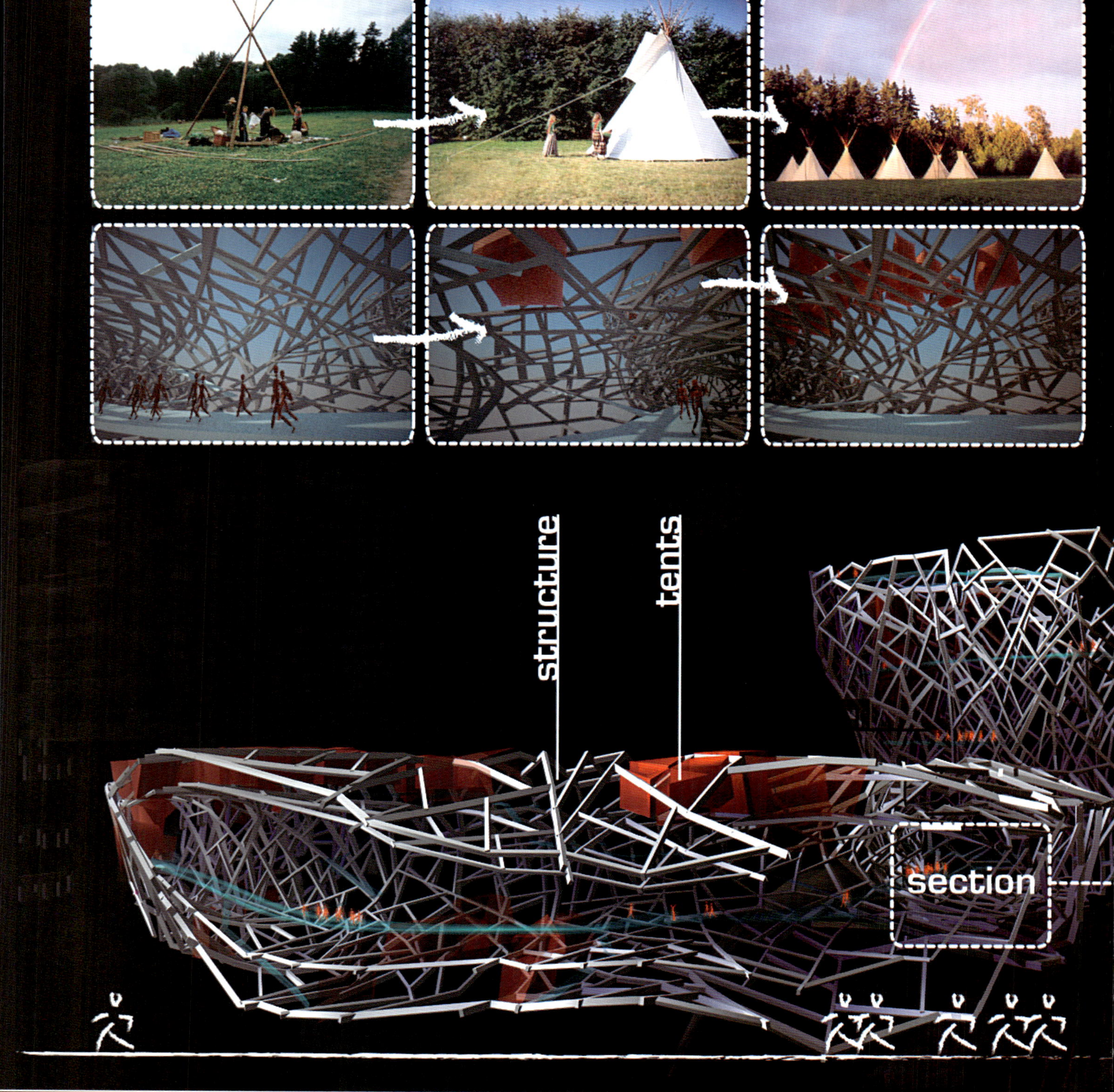
structure
tents
section

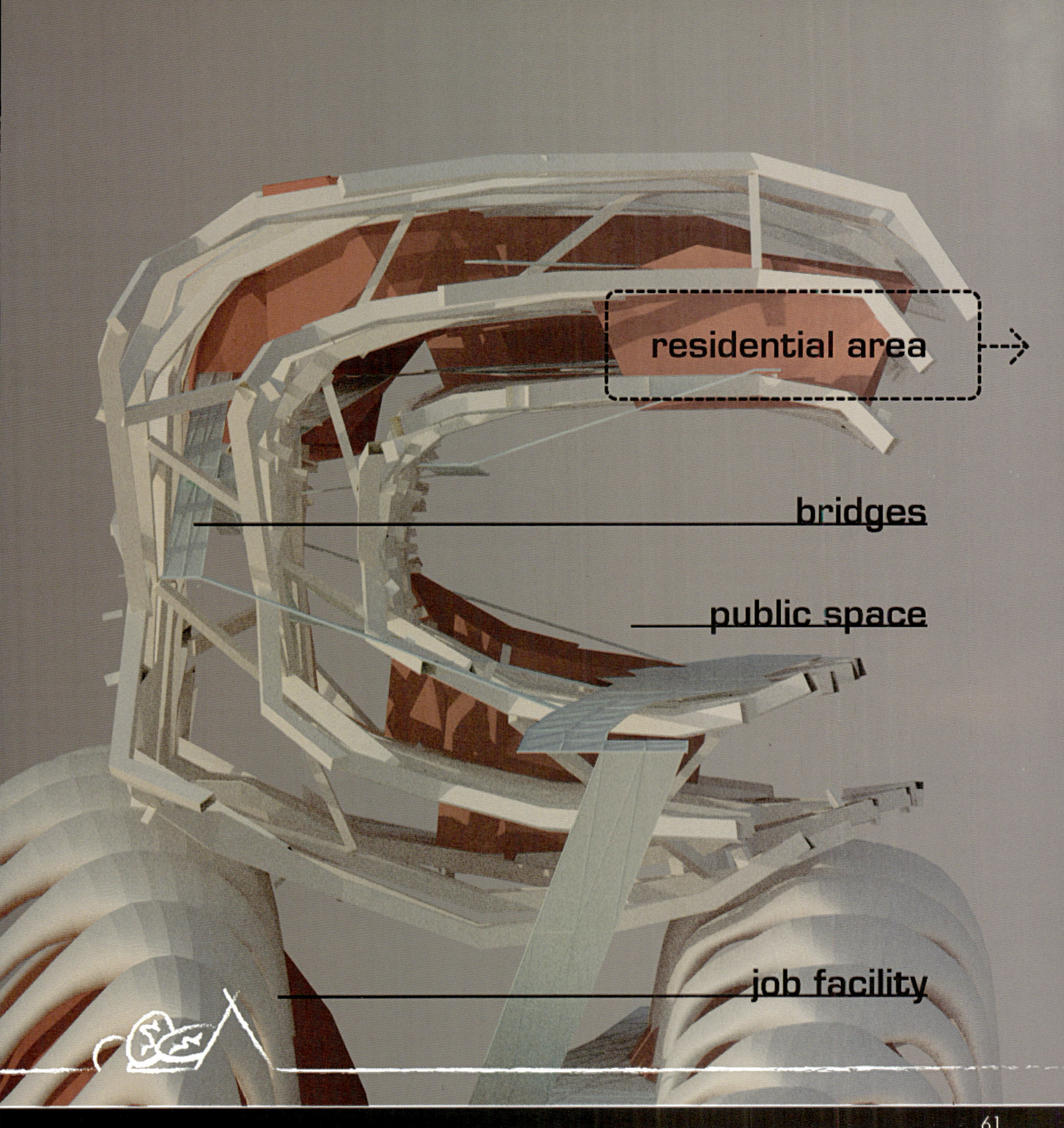
residential area
bridges
public space
job facility

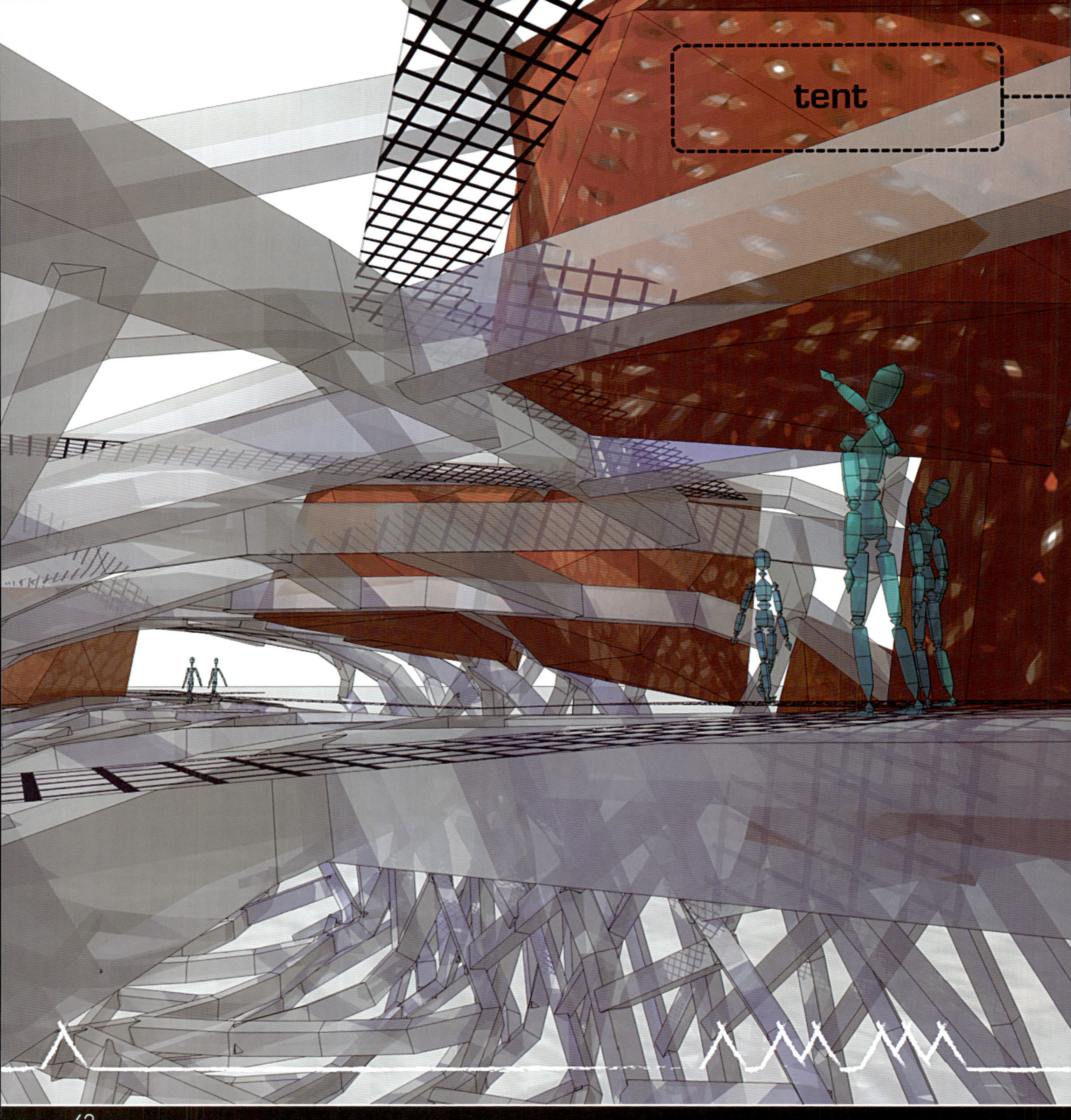
tent

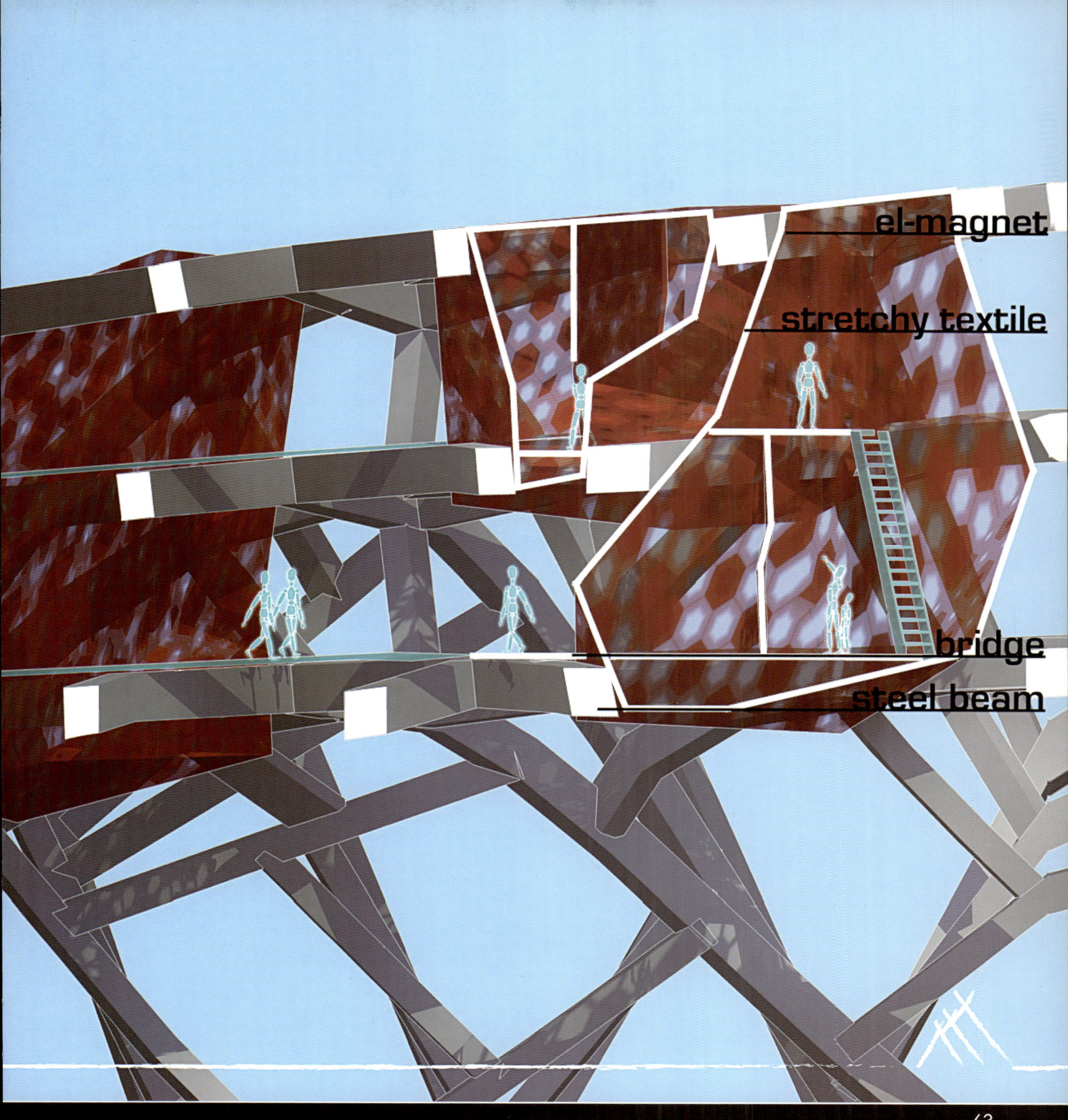
el-magnet
stretchy textile
bridge
steel beam

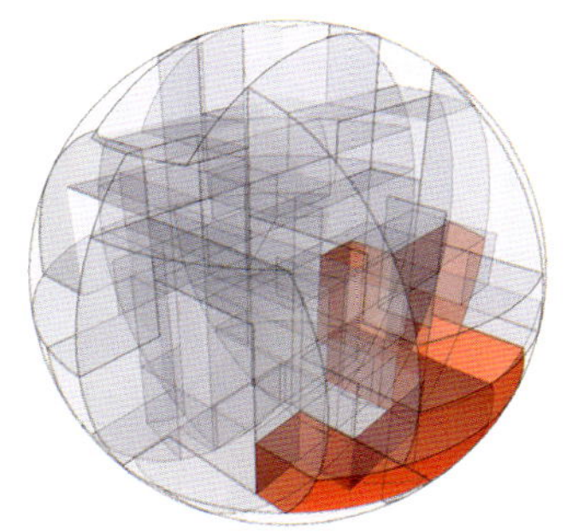

Zone 9 - Hexagonic
六角形
Alexander Bauman

与20多人一起分区完成一个大型项目的设计任务是很特别的，每个人都依赖于其相邻地块。这种依赖产生了一个复杂的可变系统，于是我构建了一个三维结构，它可适应各种不同需要。

它是一个基于简单规则，在垂直和水平方向上无止境添加的系统。该系统可在球体范围内组织空间。通过用参数定义基本元素（六角形），可建立一个能对外部和内部影响作出反应的巨大复杂的三维结构。

该结构可承担荷载（本身和相邻地块的荷载），用作基础设施，并限定不同功能的体量，比如博物馆或者居住。根据该系统的规则，可产生多种几何形体。设计表达的只是其中一种可能性。

The task of working with more than 20 people on one big project, divided into zones, is very special. Everybody depends on their neighbors. These dependencies create a complex system of alteration, therefore I developed a 3-dimensional structure which is able to adapt to a big variety of requirements.

It's an additive and endless system in vertical and horizontal direction based on simple rules. The system can organize space within the sphere. By defining the basic element (the hexagon) in parameters, it's possible to create a huge and complex 3-dimensional structure which can react to external and internal influences.

The structure is able to carry loads (itself and loads of the neighbors), is used as infrastructure and it defines volumes for functions, like museums or dwellings. Following the rules of the system it's possible to generate multiple geometries. The shown design is only one possibility.

System

modulation 1

modulation 2

3D Print

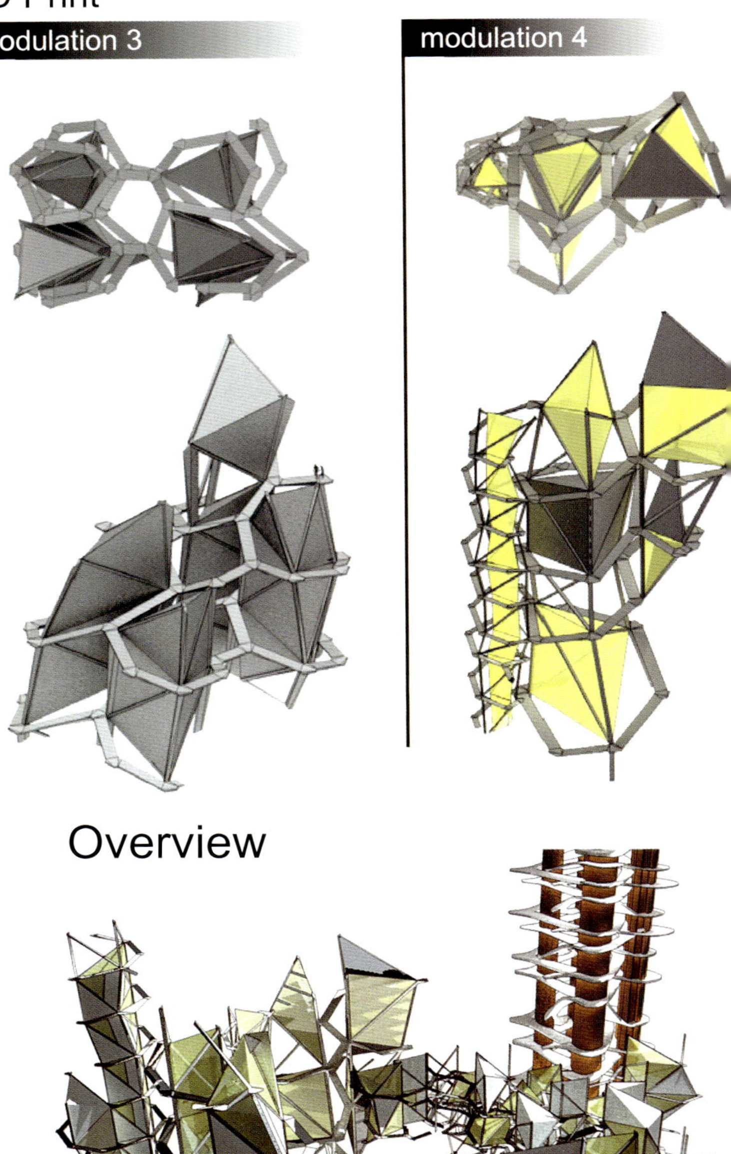

Overview

Tower 01 + 02

art studios
art museum II
skybar
art museum I
elevator
theatre

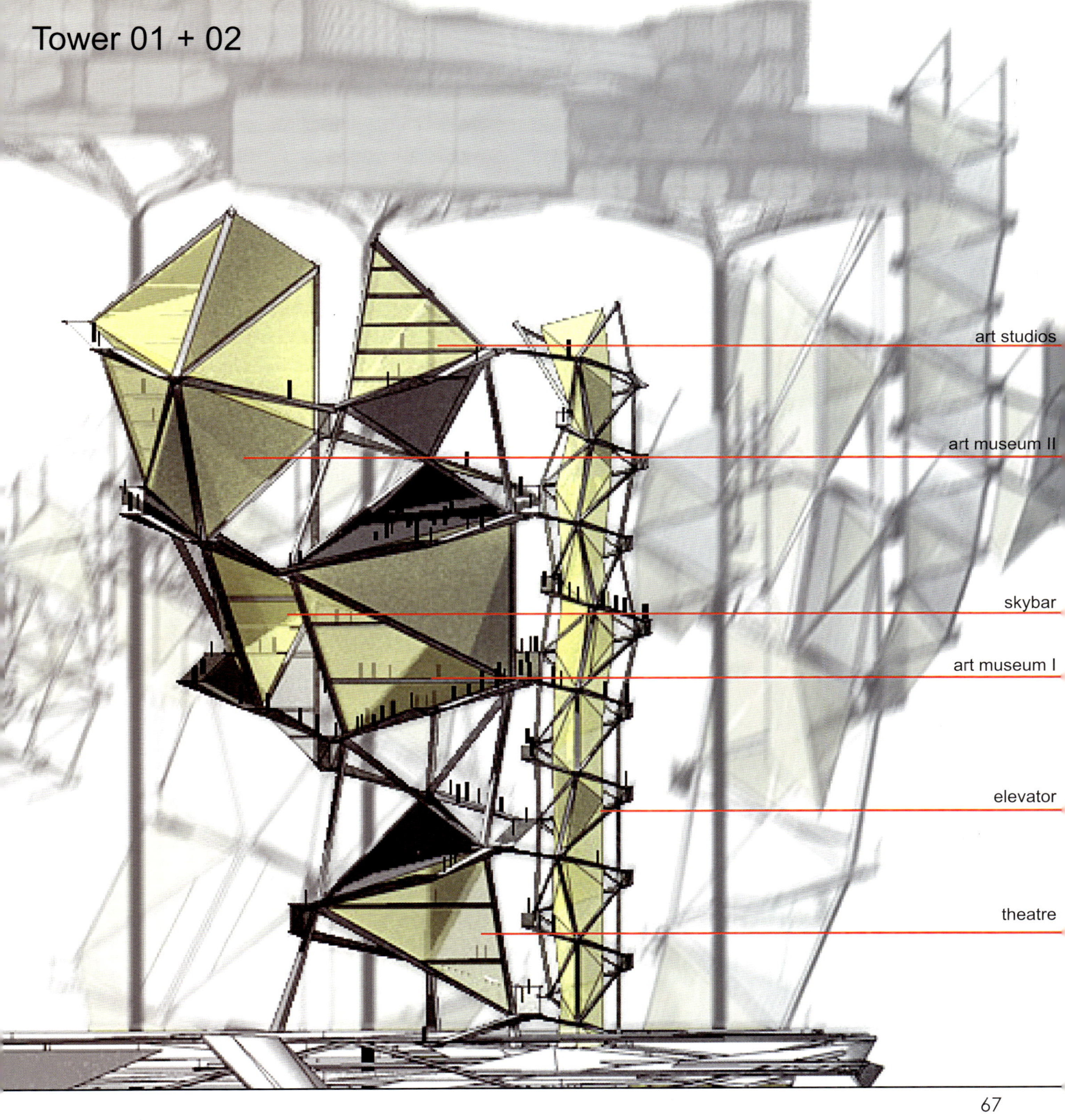

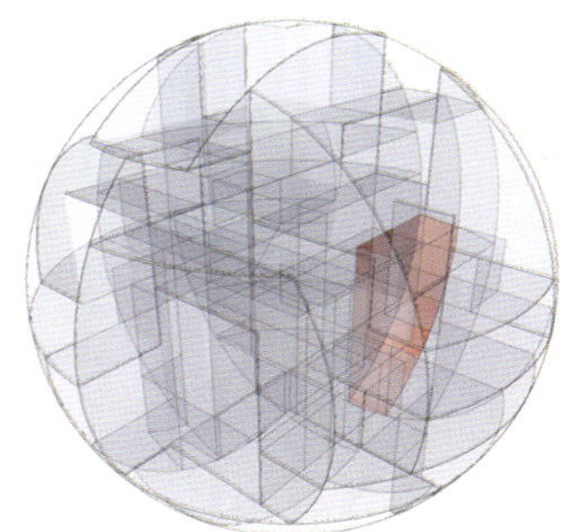

Zone 10 - 3D City
三维城市
Jedrzej Kolesinski

第10区是我对北京的重新诠释，我试图找到如何在三维地块中设计的方法。该想法是重新利用北京胡同及其不同元素间关系的参数，并将其整合为一个设计。整个区域本身是可持续的，因为利用了地热和沼气能量。整个三维模型同样是参数化的。

Zone10 is my reinterpretation of Beijing and trying to find out solution how to design in a 3D plot. The idea is to reuse parameters of Beijing Hutongs and relations between their different elements and merge them into a design. The whole zone by using geothermal and biogaz energy is self sustainable. Also the whole 3D model is parametric.

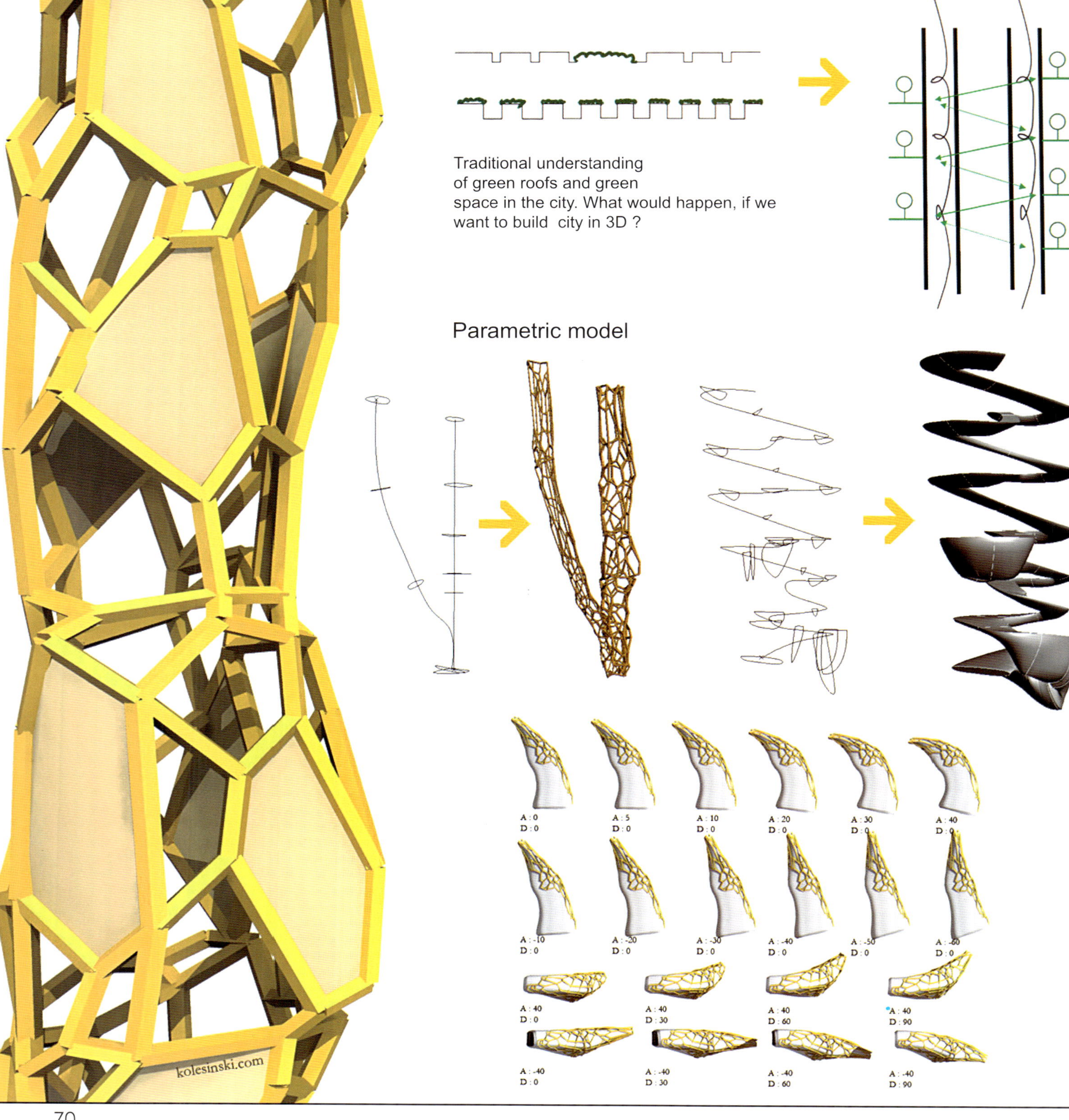
Traditional understanding of green roofs and green space in the city. What would happen, if we want to build city in 3D ?
Parametric model
A : 0 D : 0
A : 5 D : 0
A : 10 D : 0
A : 20 D : 0
A : 30 D : 0
A : 40 D : 0
A : -10 D : 0
A : -20 D : 0
A : -30 D : 0
A : -40 D : 0
A : -50 D : 0
A : -60 D : 0
A : 40 D : 0
A : 40 D : 30
A : 40 D : 60
A : 40 D : 90
A : -40 D : 0
A : -40 D : 30
A : -40 D : 60
A : -40 D : 90
kolesinski.com

Commercial space. Most of commercial space is organized as a market place. At the evening the space is converted to public space.

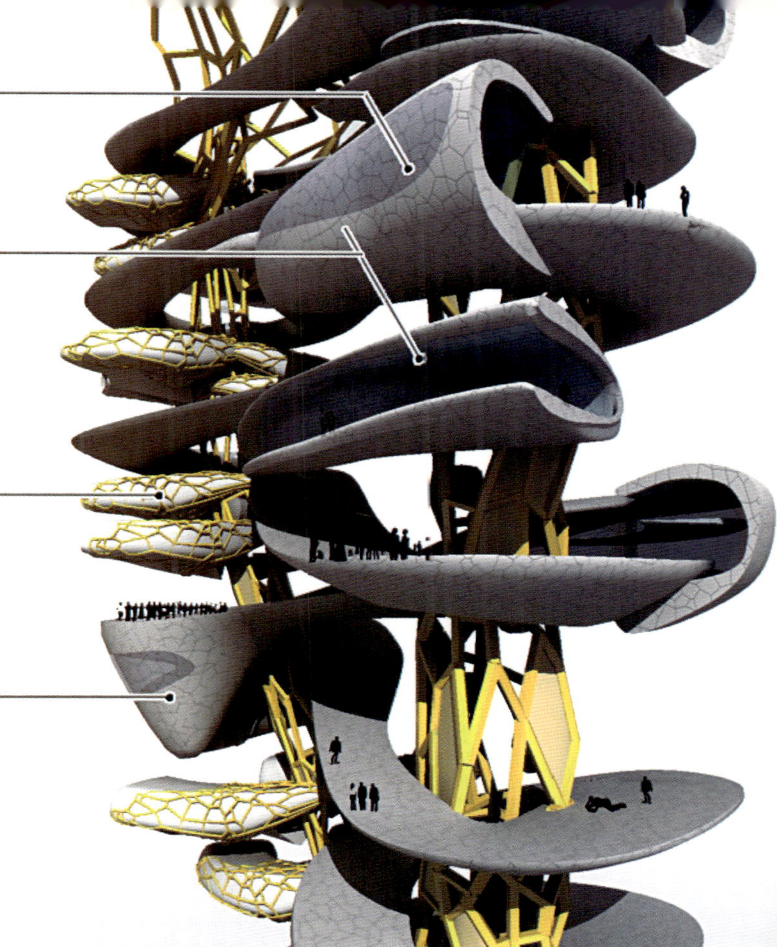

Office space. Office space is reserved for social and cultural related organizations. At least 30% of office space is for company incubators. 20% of all office volume space is used by normal offices.

Housing. Offers new quality to Beijing citizens. Each apartment unit has its own view to the Beijing panorama, however keeping values like private courtyard and individuality. Every housing unit is bit different from the others due to parametric design.

Culture. Performance scene for young Chinese artists. The space would also used for traditional Beijing opera. During breaks visitors can use Sky Bars with view to Beijing.

Multimedia eyes. Because of the fact that plot is on of only few entrances to the sphere on m-eyes users can see in real time what they can do in other plots. Other plots (plots owners) could buy m-eye time on display to promote their functions. When the user gets closer to the m-eye, he/she can decide which plot can be shown on the screen. When there are important events e.g. Olympic Games Beijing 2008 m-eye gathers people in front of it to watch those events.

Statistics

Residential	: 5110 m^3
Office	: 12455 m^3
Commercial	: 7345 m^3
leisure & culture	: 5110 m^3
Parking	: 5840 m^3

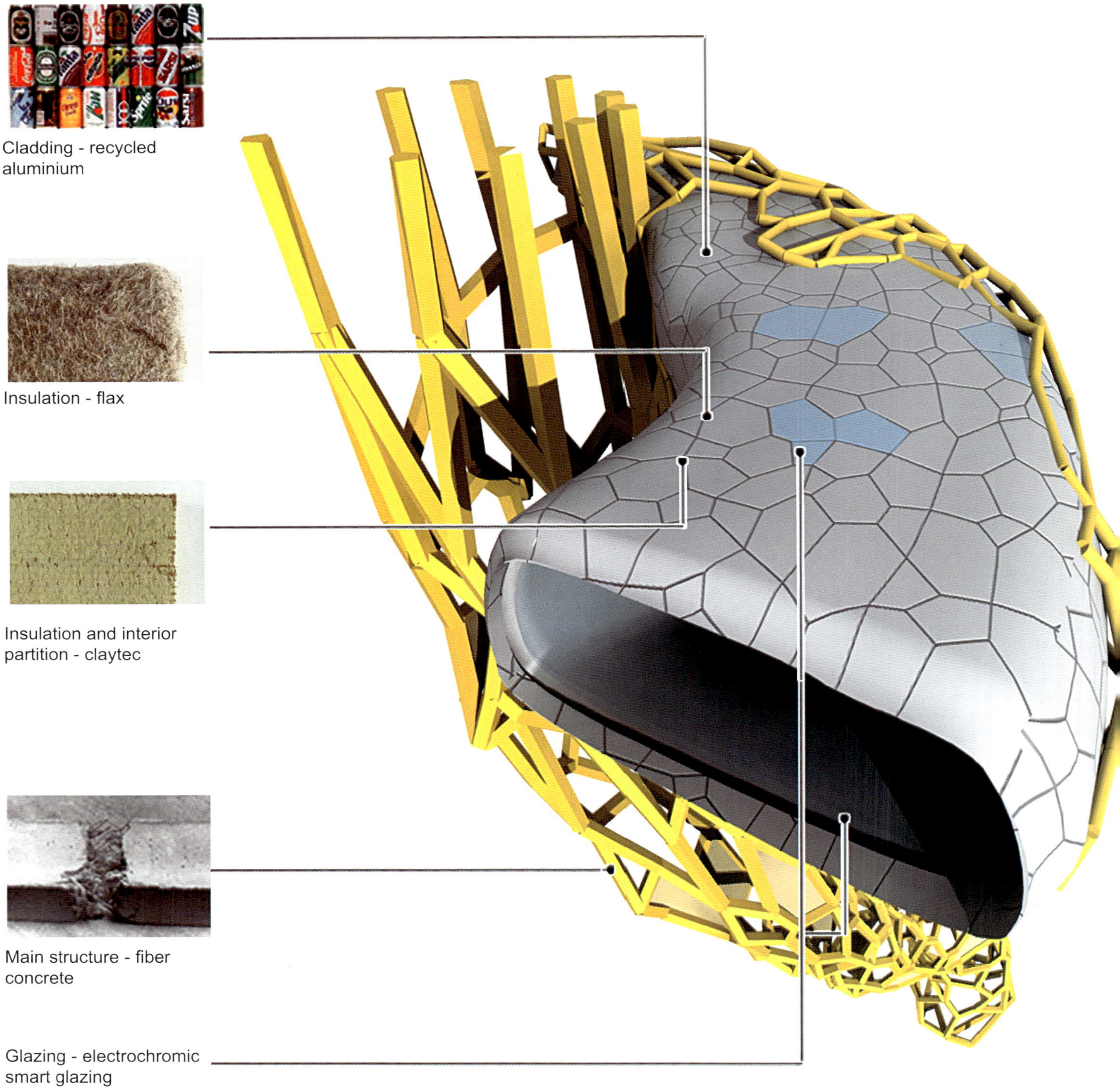

Cladding - recycled aluminium

Insulation - flax

Insulation and interior partition - claytec

Main structure - fiber concrete

Glazing - electrochromic smart glazing

Zone 9

Zone 8

Connections

Zone 11

Zone 5

Zone 18

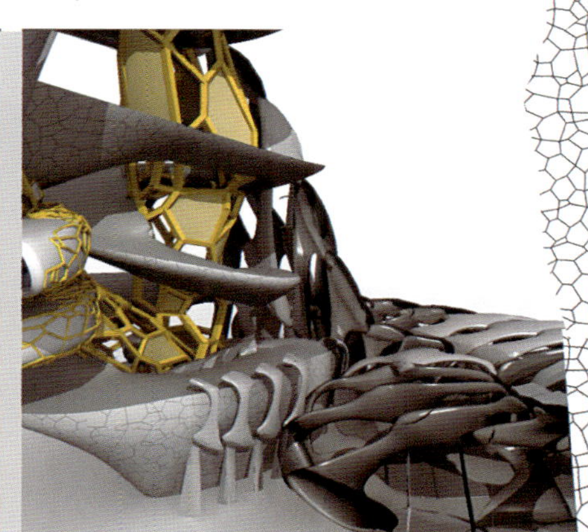

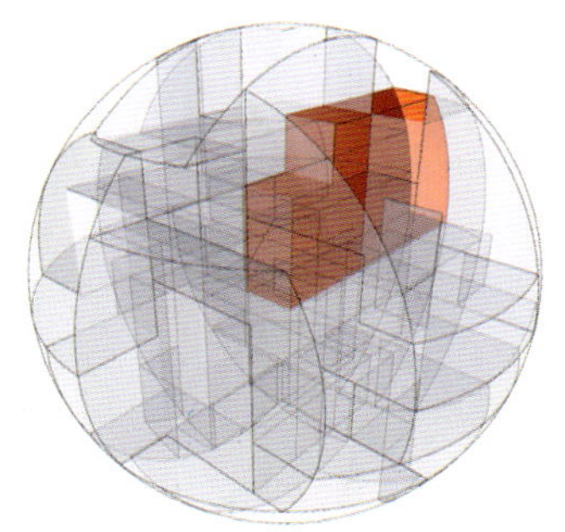

Zone 11 - Generating751
产生751

Ewoud Ruifrok

在该项目中，我开发了一个工具，可处理与相邻地块的连接和给定的功能。

该工具在virtools中创建，是个交互设计环境。它使用功能安排（立方米）、建筑概念（造型）以及设计与其环境的连接作为输入。通过设定特定的、装配和关系规则，这些输入转化为可在设计的进一步发展中使用的输出。

该工具可让我在贯穿设计过程、不断变化着的邻里环境中迅速改变设计，产生新的方案。该过程在这种产生的可能性和建筑师作为设计者的角色等方面有很多问题。

该项目围绕一个螺旋核建造，与相邻地块的连接即从该螺旋核建立起来。由该工具产生的楼面布置形成了最终设计。楼面可根据其功能通过敞开与围合而变得私密或公共。

In the project I made a tool that allowed me deal with the connections to my neighbours and the given program.

The tool is made in virtools a 3D interactive design environment. It uses the functional program (m^3) architectural concept (shape) and connections of the design to its surroundings as input. By setting special, constructional and relation rules this input is converted into an output that can be used in the further development of the design.

The tool allowed me to quickly change my design and generate new layouts in a changing environment of neighbours that are going through the design process. This process holds multiple questions about the possibility of this kind of generating and the role of an architect as a designer.

The project is build around a spiralling core from which connections to neighbouring plots are established. The final design is created by the floor layout that was generated by the tool. Floors can be made private or public by opening and closing them depending on the function they hold.

input	processing	output

shape

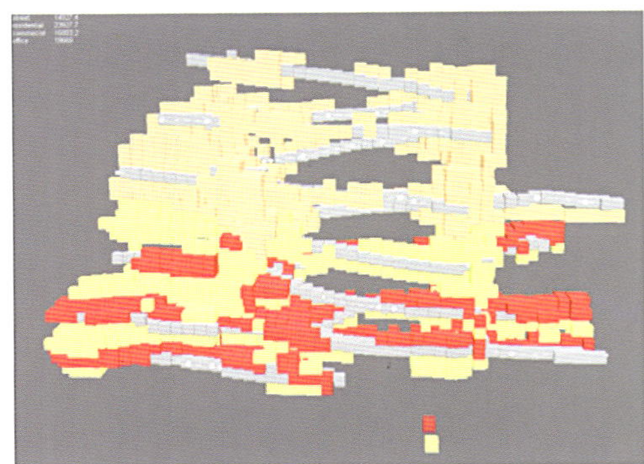

floors

Add Column Add Row

	0 : curve	1 : function	2 : max size	3 : group	4 : 2e function
0	New Curve.0002	commercial	6800.0000	curve2	office
1	New Object	commercial	16000.0000	curve1	office
2	New Curve.0003	commercial	6000.0000	curve3	office
3	New Curve.0004	commercial	5500.0000	curve4	office
4	New Curve.0005	commercial	3300.0000	curve5	office
5	New Curve.0006	commercial	4800.0000	curve6	office
6	New Curve.0007	office	5100.0000	curve7	office
7	New Curve.0008	resi	1200.0000	curve8	resi
8	New Curve.0009	resi	3200.0000	curve9	resi
9	New Curve.0010	resi	3500.0000	curve10	resi
10	New Curve.0011	resi	2700.0000	curve11	resi
11	New Curve.0012	resi	1700.0000	curve12	resi
12	New Curve.0013	resi	1800.0000	curve13	resi
13	New Curve.0014	resi	1000.0000	curve14	resi
14	New Curve.0015	office	800.0000	curve15	office
15	New Curve.0016	office	1900.0000	curve16	office
15	New Curve.0017	office	2250.0000	curve17	office

function

skin

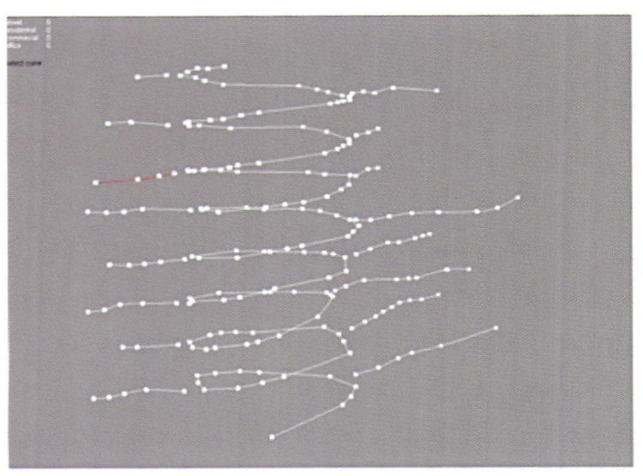

curves

facade

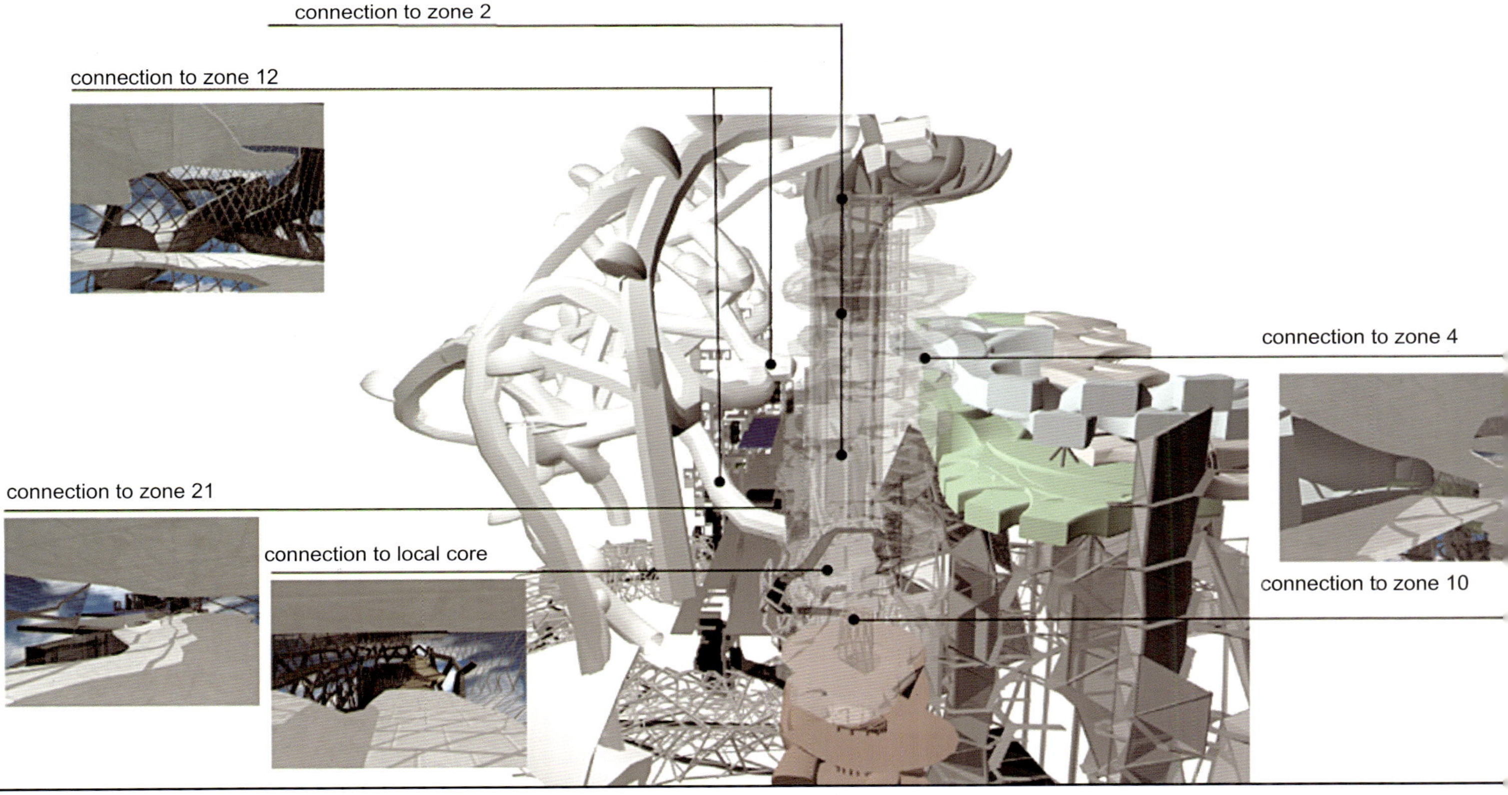

connection to zone 2

connection to zone 12

connection to zone 4

connection to zone 21

connection to local core

connection to zone 10

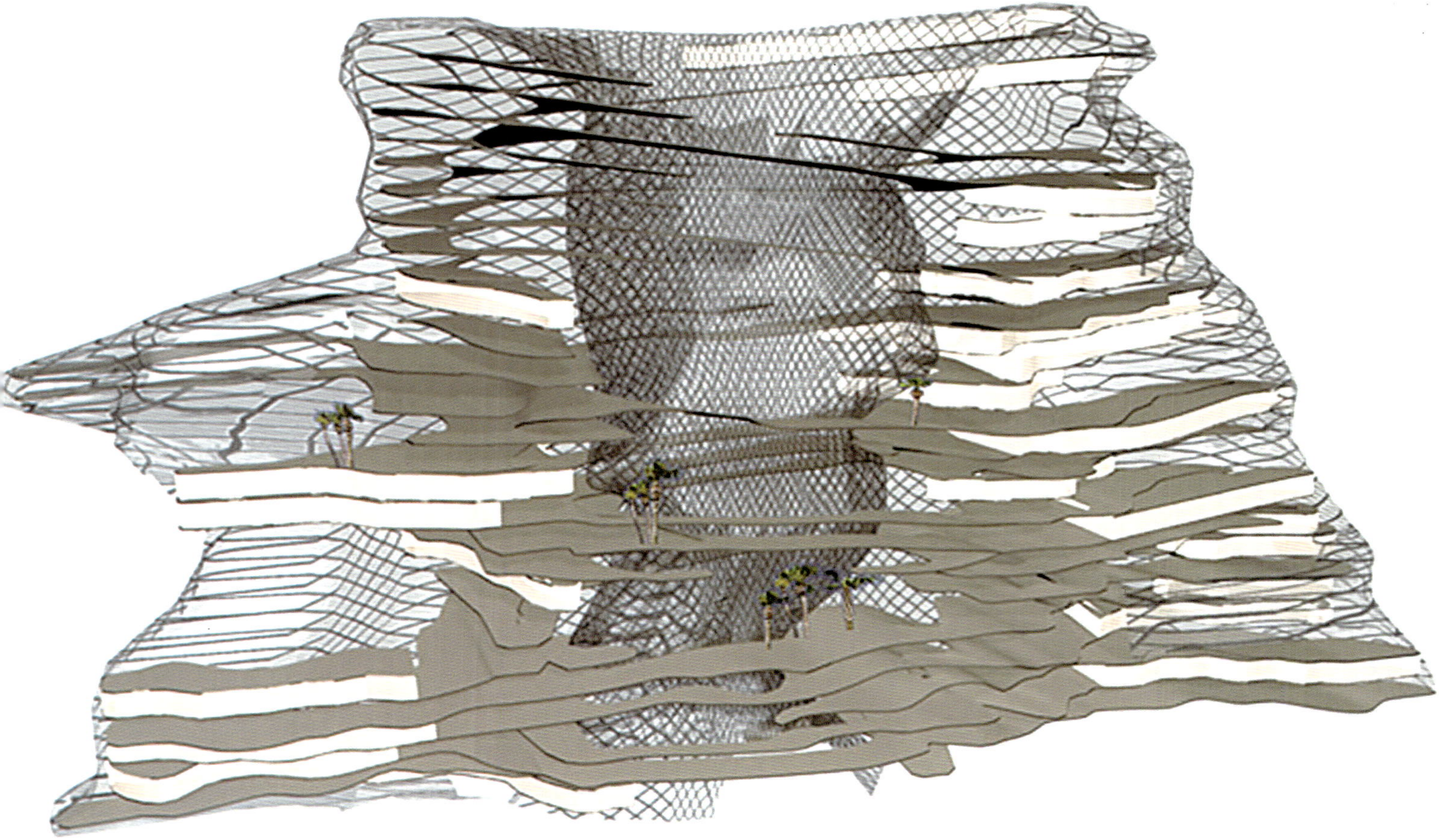

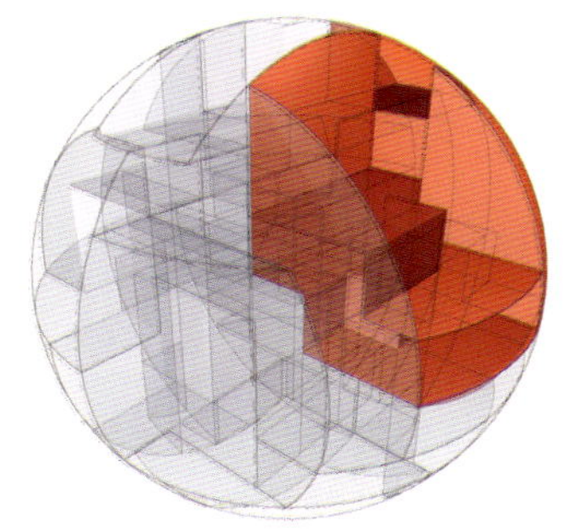

Zones 12-15 - The Hive
蜂巢

Jenna Fizel, Eva Kiesel, Isil Sencar, Felix Wurst

蜂巢是一个活着并成长着的有机城市。蜂巢中的生命出现于一个预置的由主要结构和交通纹理组成的有机体，它开始成长并根据居民的要求而变化。

蜂巢由一个作为结构骨架的交通系统组成，空间附着于或产生于这个结构，并被一层外皮包裹。交通系统在三个层面上运作：传送人、单元和工具。外皮是蜂巢的遮盖装置。它保护蜂巢防止外部因素侵袭，赋予其个性化外观，提供可控的空间，并包括了大多数生态方面的内容。空间被分为两组：移动单元（居住、办公、商业）和固定单元（工厂、凉棚、研究所、影剧院）。

各单元基于参数来设计，这些参数由设计者构建，并由居民进行选择。他们可在工厂生产。工厂是作为蜂巢的大脑，因为它通过三项主要任务向整个地块分配信息和空间：建立/重复利用单元，调节运动和维护支持系统。蜂巢中的其他关键区域有博物馆、影剧院、研究与设计机构和凉棚。

The hive is an organic city that lives and grows. From a preset organism consisting of main structural and transportational veins, the life in the hive emerges and it starts to grow and change according to the demands of the inhabitants.

The hive is composed of a transportation system which acts as the structural skeleton and spaces attached to or emerging from that structure, enveloped by a skin. Transport system works in three layers, transporting people, units and utilities. Skin is the covering mechanism of the hive. It protects the hive from external factors, gives its characteristic appearance, supplies with controlled spaces, involves most of the ecological aspects. Spaces are divided into two groups: moving units (residential, office, commercial) & fixed spaces (factory, arborteum, institute, theatre).

Units are designed based on parameters created by designers and preferences chosen by the inhabitants. They can be manufactured at the factory. Factory acts as the brain of the hive as it distributes information and space throughout the plots with three main tasks which are creating/recycling units, regulating movement and maintaining the support system. Other key areas in the hive are the museum, theatre, research and design institute and Arborteum.

myspace music

Click here!

myhivemusic Tune in Now to check out the Top Artists

kimkong

Male
28 years old

Beijing

Last Login:
11/30/2006

View My: Pics | Videos

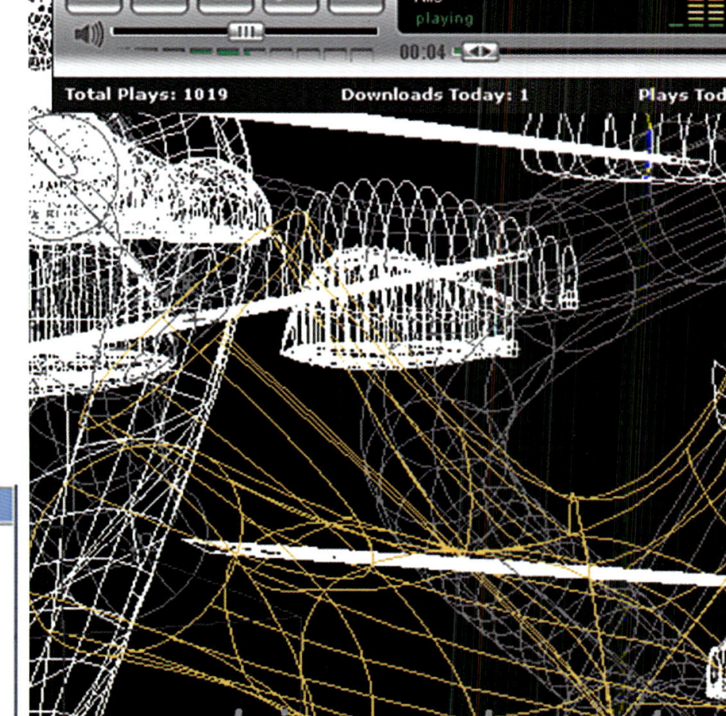

Contacting kimkong

Send Message | Forward to Friend

Add to Friends | Add to Favorites

Instant Message | Block User

Add to Group | Rank User

your hive network:

räuberhöhle

Nina Hemmingsson

Gaffaman

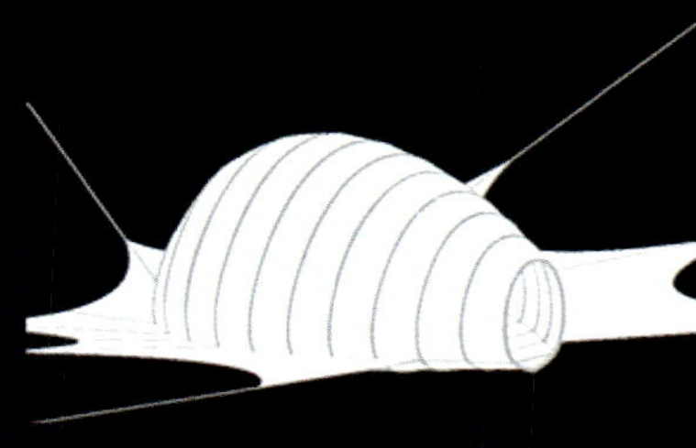

NORA BELOW

sara olausson

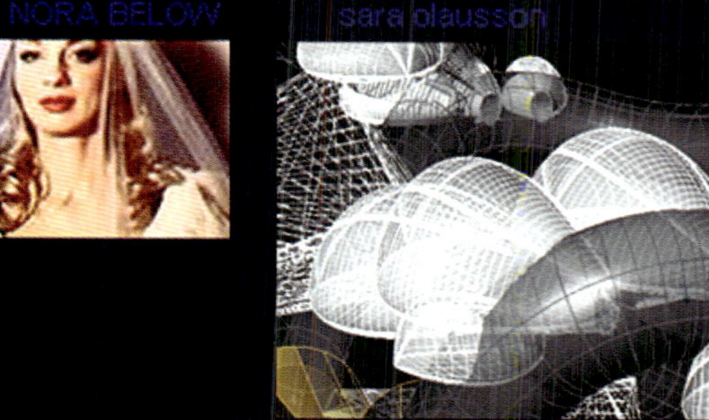

DATA EXCHANGE

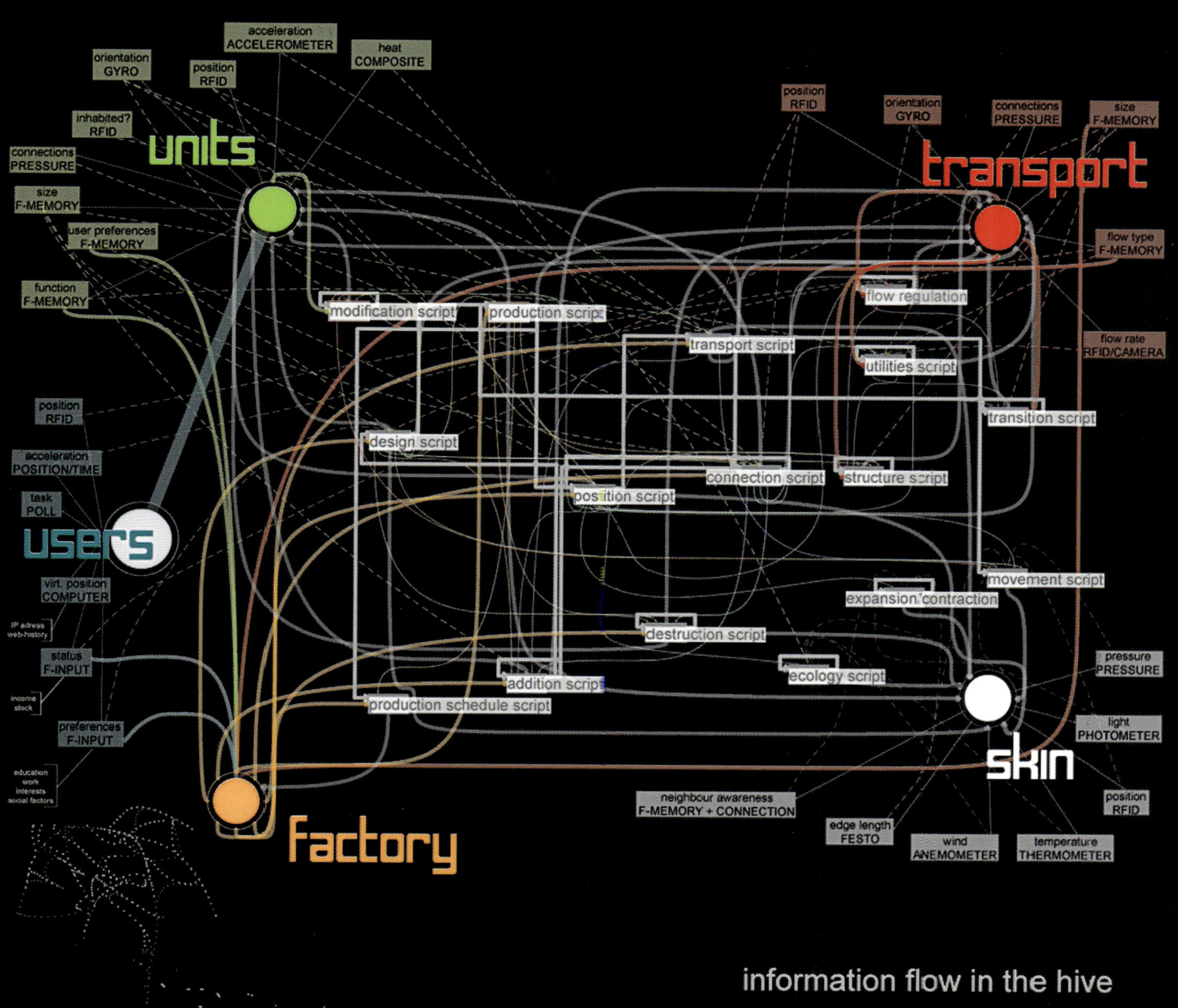

information flow in the hive

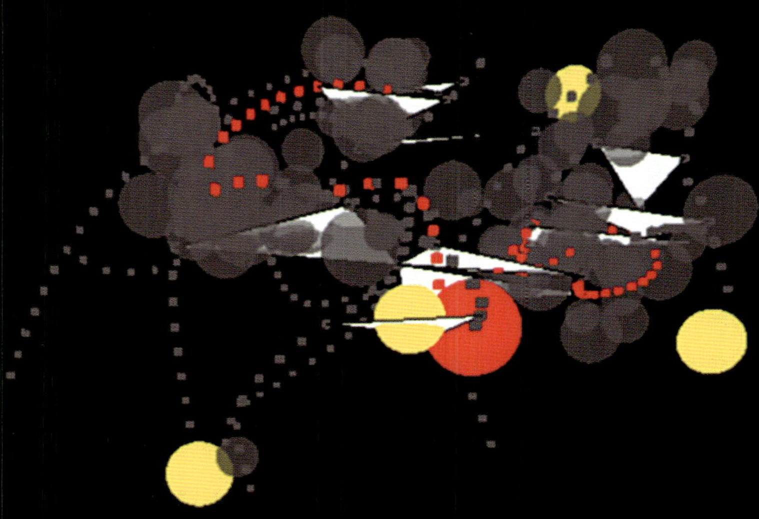

trace class

in order to find the minimum distance between a unit's position and a set of attractors, the distance, along the most efficient transport line to each attractor must be found, summed, and compared to that at each other available position.

it is inefficient for each unit to have to calculate this distance itself.

instead, each position is given knowledge about its own distance from all attractors.

this information need only be gathered once, but can be queried by any unit/plane at any time

plane class

each plane attaches itself to three points.

it's represented by a triangle of the same general shape.

the plane is constrained with values such as maximum slope, minimum and maximum edge length, and minimum height (triangle), specified in the code.

unit class

each unit attaches itself to a point.

it's represented by a sphere of the unit's maximum diameter.

this sphere collides with other units, planes, and the tranport particles to ensure that the unit does not intersect anything.

FACTORY

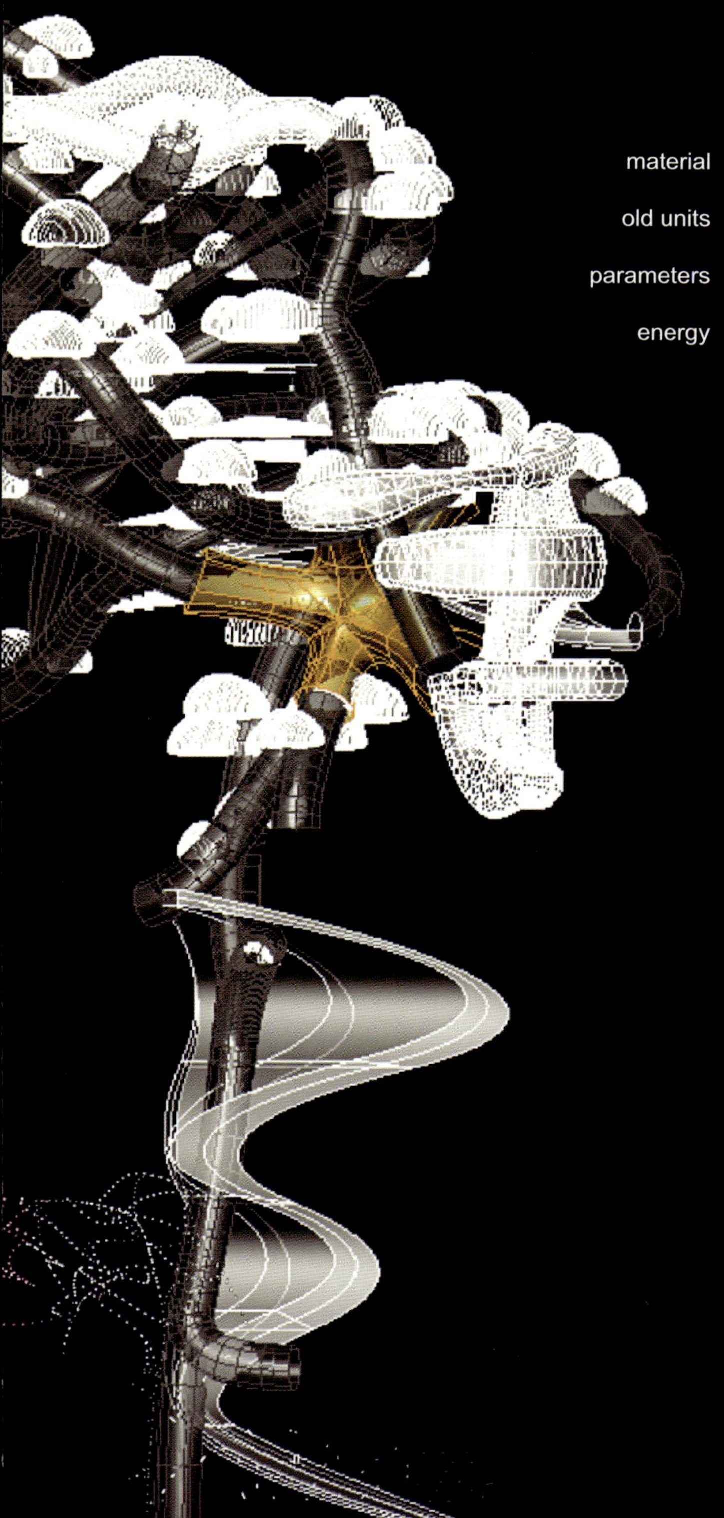

material
old units
parameters
energy

new units
repaired skin
local transport sections
position information

three physical tasks of the factory

create/recycle units

uses design parameters to produce space based on user preference

regulate movment/placement of units

uses relationship parameters to determine location of all units in the system, determines when change is appropriate

maintain support systems

has responsibility for upkeep and modification of transport and skin systems, ensures that all systems grow and shrink appropriately

UNITS

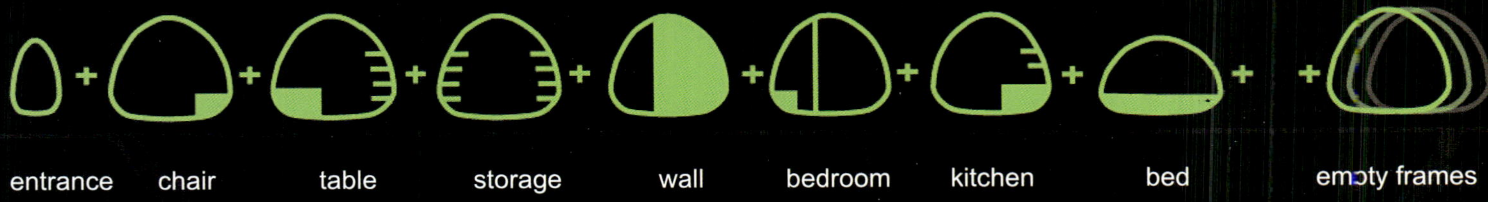

entrance + chair + table + storage + wall + bedroom + kitchen + bed + empty frames

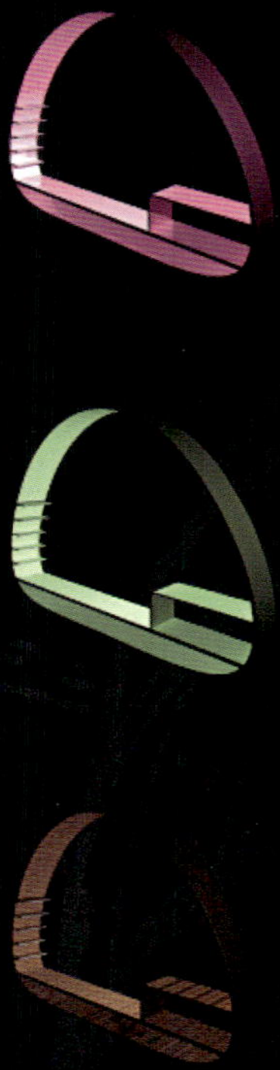

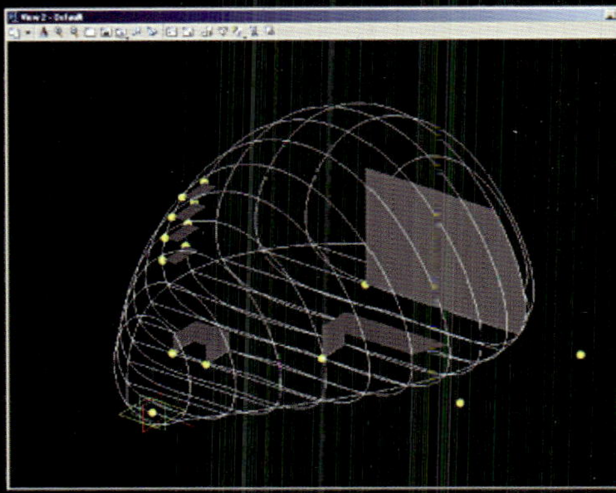

design is influenced by the owner:
volume furnishing colour materials

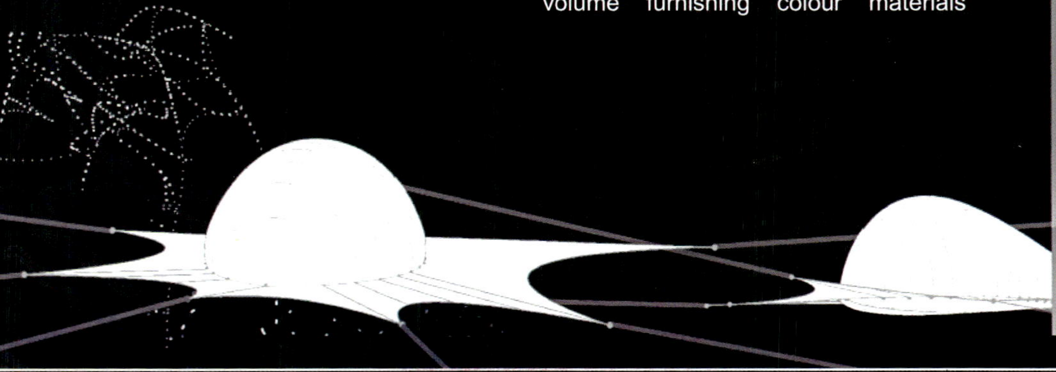

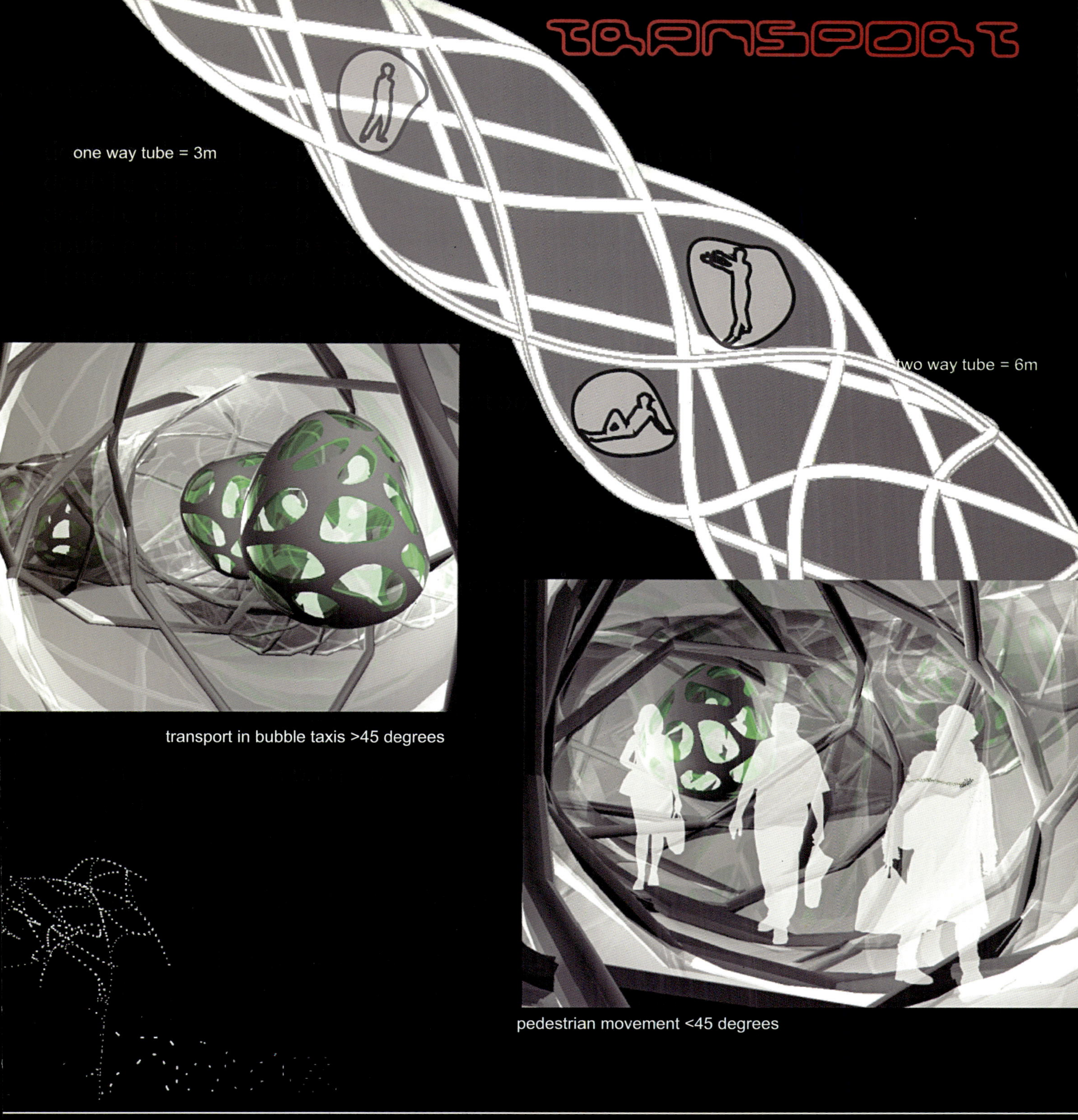

transport in bubble taxis >45 degrees

pedestrian movement <45 degrees

skin

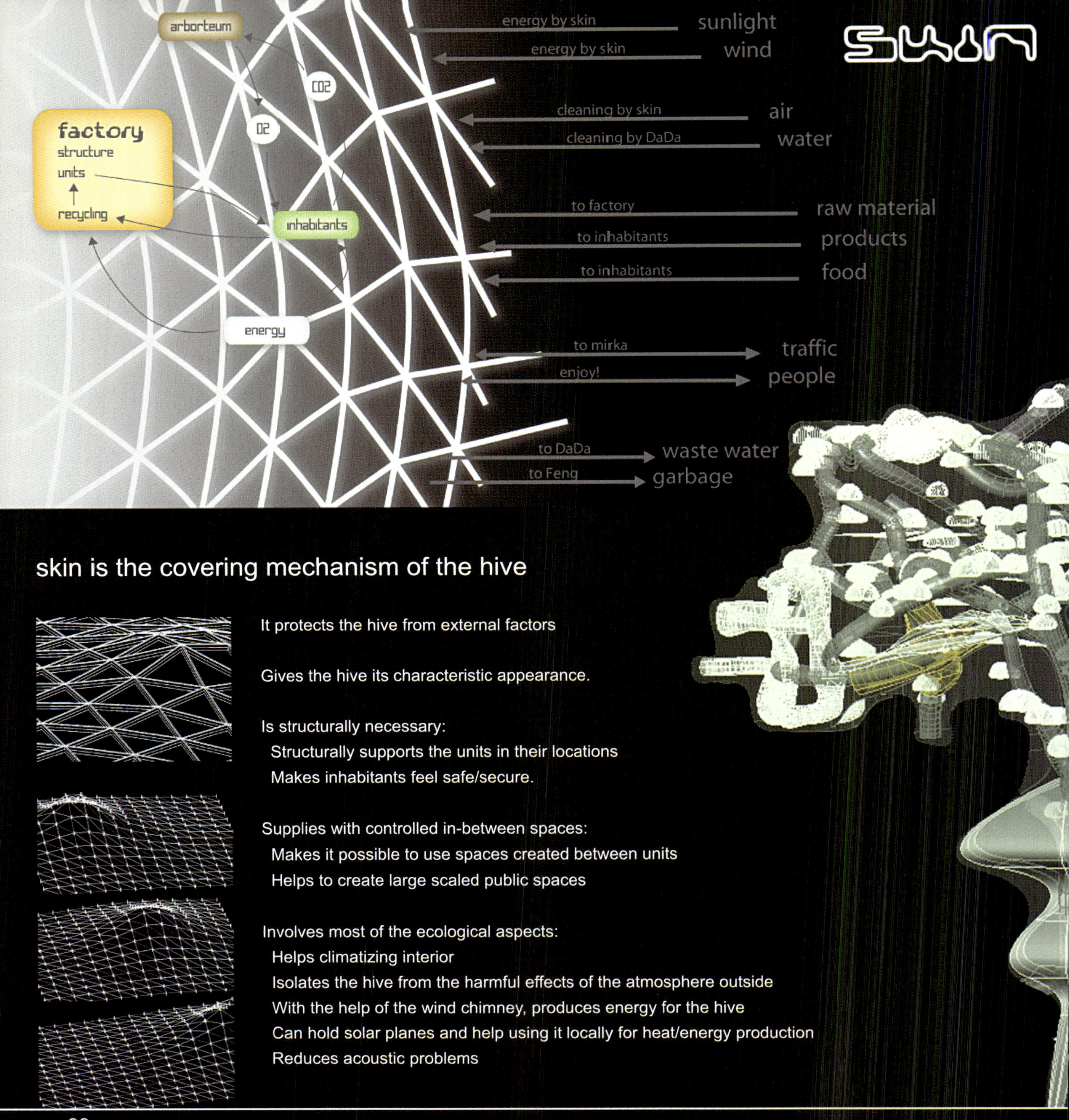

skin is the covering mechanism of the hive

It protects the hive from external factors

Gives the hive its characteristic appearance.

Is structurally necessary:
- Structurally supports the units in their locations
- Makes inhabitants feel safe/secure.

Supplies with controlled in-between spaces:
- Makes it possible to use spaces created between units
- Helps to create large scaled public spaces

Involves most of the ecological aspects:
- Helps climatizing interior
- Isolates the hive from the harmful effects of the atmosphere outside
- With the help of the wind chimney, produces energy for the hive
- Can hold solar planes and help using it locally for heat/energy production
- Reduces acoustic problems

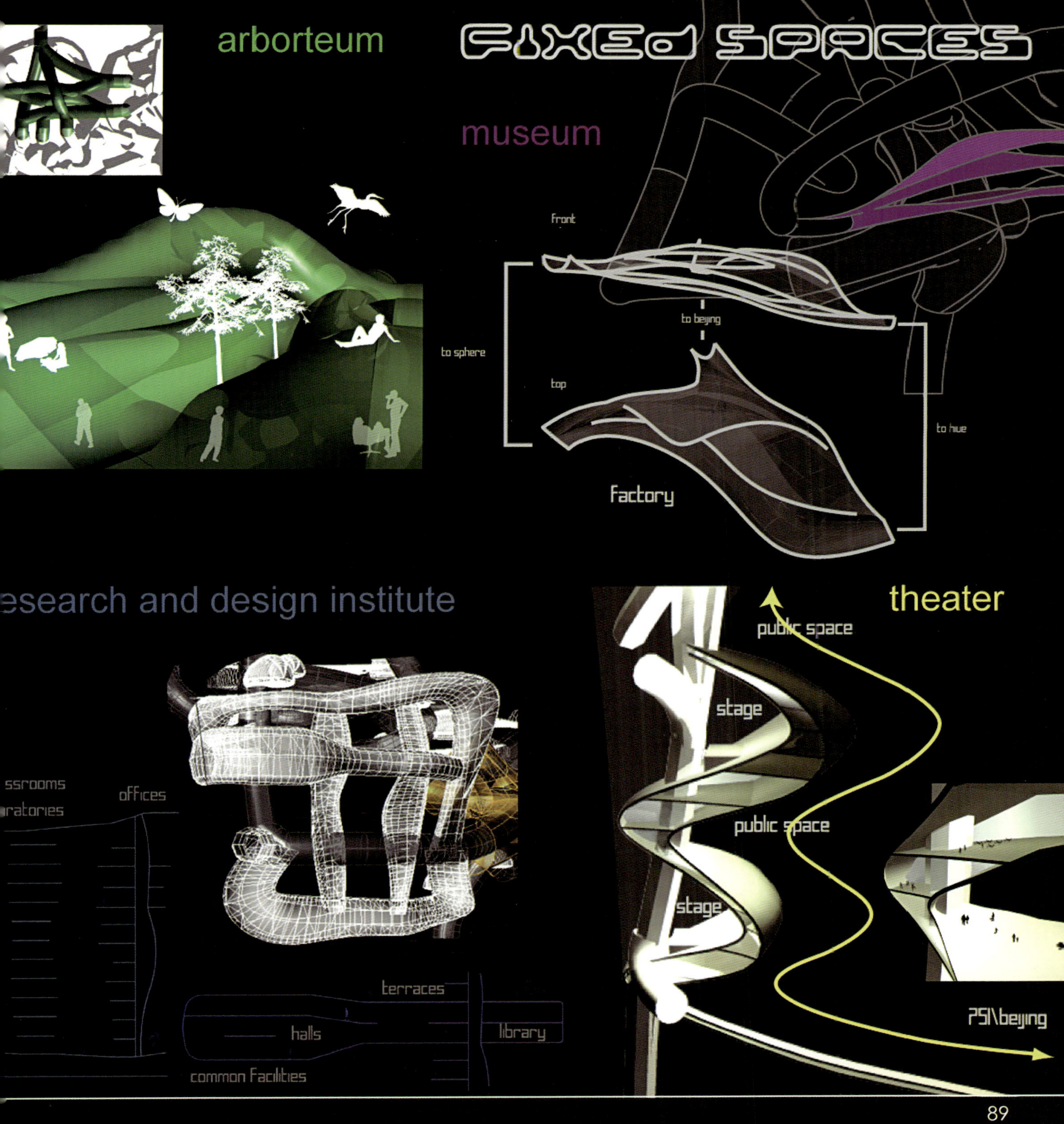
arborteum
FIXEd SPACES
museum
Front
to beijing
to sphere
top
to hive
Factory
esearch and design institute
theater
public space
stage
public space
stage
ssrooms
offices
ratories
terraces
halls
library
common Facilities
PSI\beijing

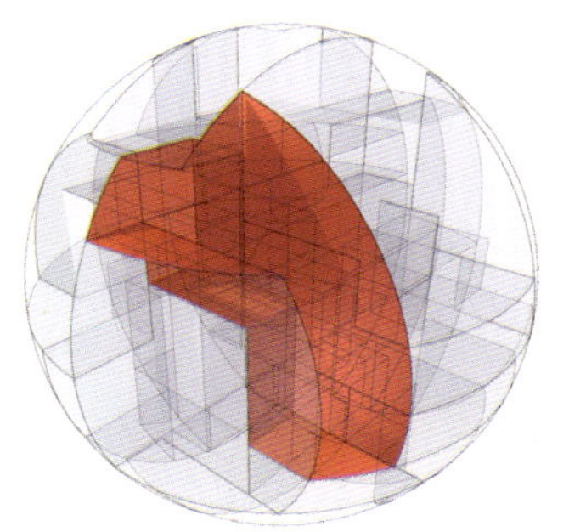

Zone 16 - House+Office
住宅+办公
Magdalena Garbarczyk

该项目的目标首先是通过创造一种功能混合的空间，让人们可以在一个地方附近拥有所有必要的功能，来解决北京不断增加的私人交通问题（实际上北京有200万辆小汽车）。

将所有这些功能细分为小的模块单元或容器的想法，来自北京798和751区域的文脉。确实，这个转变为“艺术家庇护所”的工业区域，在非常动态的工业环境中给人一种不断变化的感觉。这就是为什么该项目总体概念是运动的自由。让人们不但具有移动身体的可能性，还具有移动其空间的可能性，在不同结构中游戏，在私密与公共之间转换，这些是“居住+办公”想达到的目标。

居住和办公组合在一起形成的“双胞胎单元”模块可以共享、交换、取代和自由移动，创造了一个无限变化的空间。

The first aim of this project is to bring a solution to the issue of increasing individual traffic in Beijing (where there is actually 2 million cars) by creating a space where a combination of functions gives people the possibility to stay in the area of the plot with all necessary functions nearby.

The idea of spliting all these functions into small modular units or containers is inspired by the context of the 798 and 751 districts in Beijing. Indeed, these industrial areas transformed into “artists' refuge” give the feeling of a constant change in an industrial very dynamic environment. That's why the general concept of this project is freedom of movement. Giving people the possibility to move not only their bodies but also their space, playing with different configurations and switching between private and public are the aims “H+O” wants to achieve.

A house and an office combined together forming modular “twin-units” can be shared, exchanged, replaced, freely removed creating an endlessly changing space.

concept

Two main functions organize the plot: residential and offices -considered rather as manual working spaces like ateliers than desk-and-computer configured offices. The target group is especially people who need to have their working space not only during official working hours and need to have the possibility to provide different uses of the same space. For instance, artists or craftsmen could transform their individual working spaces into common exhibition halls. The platforms which support the containers are considered as collective spaces especially used by owners of the "twin units" but become public spaces during "open-ateliers" days or exhibitions. For more free open-air exhibitions, the central and the top platforms, considered as the main public spaces of the plot, can be used. The central platform which joins the two parts of the vertical transport system is an open-air exhibition space especially used by people living in the plot but also by all visitors and inhabitants of the sphere using the common transport system.

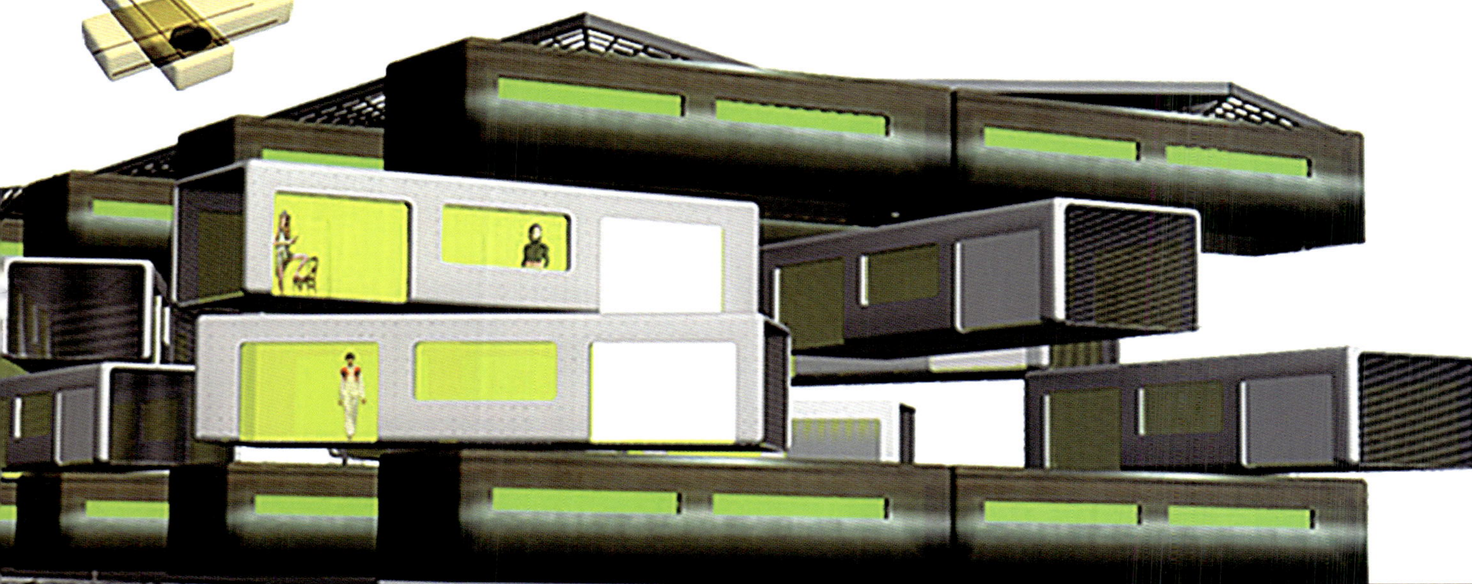

structure

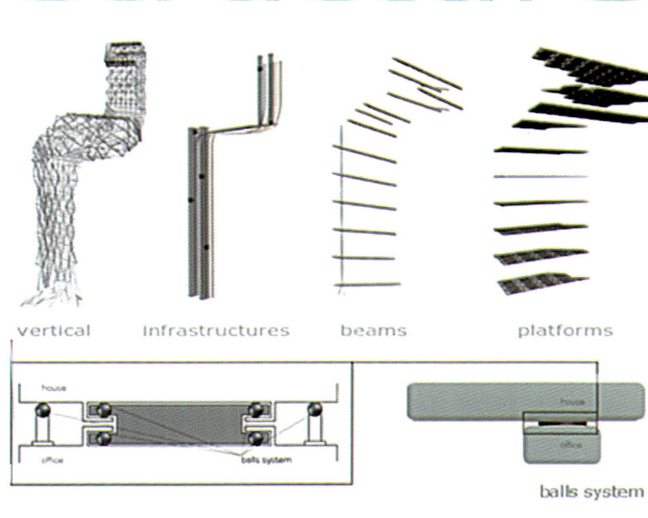

The plot is generally organized into a main steel structure system (including the vertical transport –elevators– and the infrastructures) and a lighter system, the platforms, which supports the "twin units". These "twin units" are mostly an assembly of a house and an office but people living in the plot can choose the functions of their boxes. They can also choose if they want to have them fixed or moving and if they want to connect other containers or not. The containers can be displaced using the energy of people, that's to say only by pushing them thanks to a balls system placed between two containers.

The platforms which support the containers are triangulated in order to give the possibility to remove unused ones and they are latticed to bring as much light as possible and permit the use of solar panels.

connections

zone1
using the water tower for entertainment and transport

zone2
connected by ramp

zone3
using the exhibition hall and galleries

zone8
providing vertical transport system supported by its structure

zone14
using its garden

zone15
using its school

zone17
providing vertical transport system and infrastructure going through it

zone19
connected by platforms using its cultural space

zone21
supported by its structure giving stability for its wall

zone22
connected throught zone2

The load of the structure is transferred by the beams and the column so that the point of the gravity centre is supported by plot 22 and fixed by plot 9 to keep it steady. The top platform joins plot 15's garden and is accessible to the whole sphere. It's a quiet space of distraction.

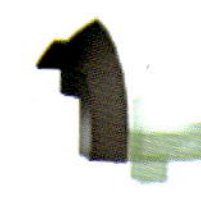

astrid/plot22 tomas/plot9 tobias/plot23 sama/pl...18 marije/plot3 julien/plot20 felix/plt15 d.../plot1 christina/plot2 isil/p...

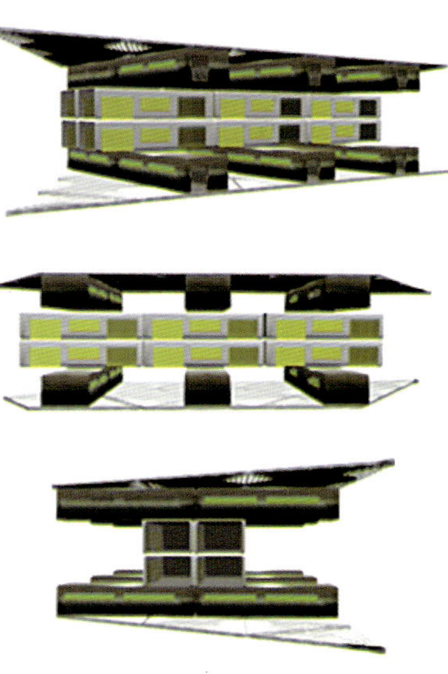

The boxes can be connected together to create exhibition spaces for artists' work. There are several possibilities of configuration. Some of them make the offices be more private without any connection with neighbors and people are free to move their boxes whenever they want. The balls system gives the possibility to collect the energy produced by friction and to bring electricity in the boxes. The energy is mainly produced by this system and solar panels placed on top of each box.

organised configuration

There are different possibilities to place the containers. The organized configuration (see upper picture) offers bigger spaces for common use. Owners need to agree on one common proposition but can always stay away from common activities by keeping their façades closed. In this case, external platforms are added in order to facilitate the routing. The free configuration (picture on the right) allows users to place their boxes according to the sun, the wind, the external conditions and to create more private spaces. This freedom of moving gives users the possibilities to hide or show, open or close, light up or darken their space.

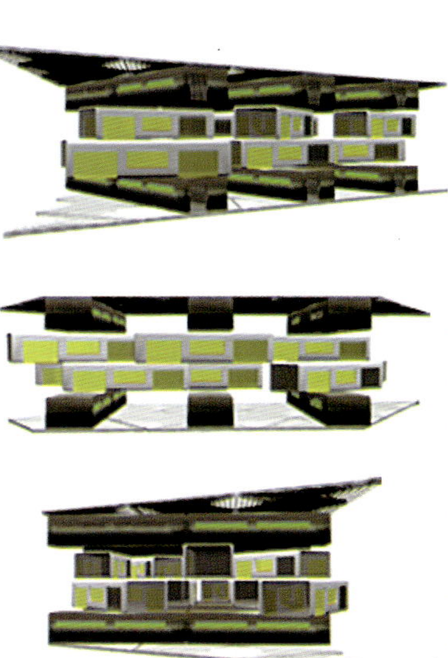

free configuration

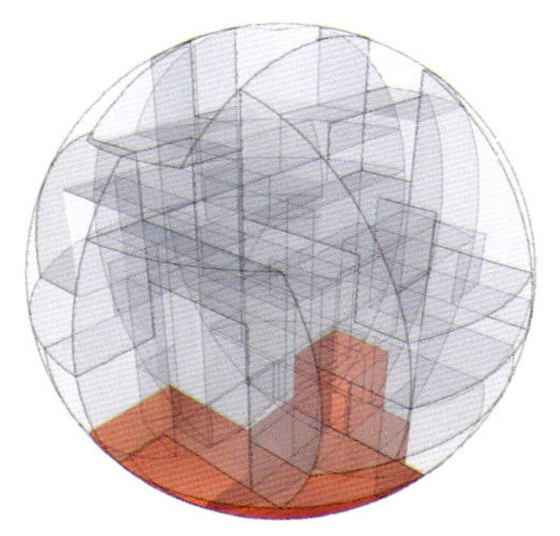

Zone 17 - City in a City
城市中的城市

Sema Alaçam Aslan

设计基地位于北京的一个名为751的老工业区。北京是个大城市。什么是城市？为什么城市的公共性很重要？什么是市民？城市中的时间意味着什么？改变时间和个性？工业时代的“小汽车”意味着什么？为何步行活动仍然很重要？

“城市一直限定着其社区，以抵御外部世界；内部有防护墙的居住区不能称为社区”。

E．E．Lozano,今日城市，1990

什么是教育的新概念呢？信息流的速度是怎样影响边界的？基础设施的连续性是目标之一，目的是：提供单元设计方法的适应性，在能量流动时减少能量消耗。教育功能包括幼儿园、小学、初中、高中和他们中的共用空间，诸如图书馆、公共课程场所。白天和晚上之间的平衡带来了空间功能的适应性。白天作为学校目的使用，晚上向公众开放课程。

The design site is located in Beijing. It is an old industrilized area called 751 District. Beijing is a metropolitan city. What is a city? Why public is important in a city? what is a citizen? What does time mean in a city? Changing time and identity? What does “car” mean in industrial age? Why pedestrian flow is still important?

“cities always define their community, as against the outside world; a settlement with internal defence walls cannot be called a true community”

E. E. Lozano, Cities Today, 1990

What are the changing concepts of education? How does speed of information flow affect borders? Continuity of infrastructure is one of the aims in order to: provide flexibility to design method of the units and decrease the energy consumption while energy flows. The function of education includes kindergarten, primary school, secondary schoool, high school and common use places among them, such as libraries, public course spaces. The balance between nght and day is providing functional flexibility to spaces. Both day usage for school purpose and courses for the open public at the night time.

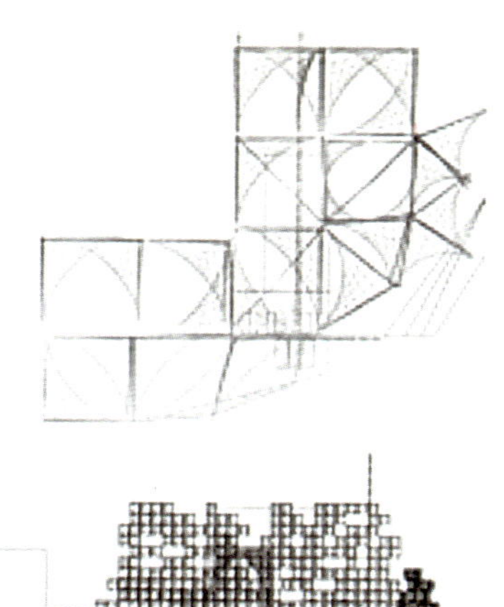

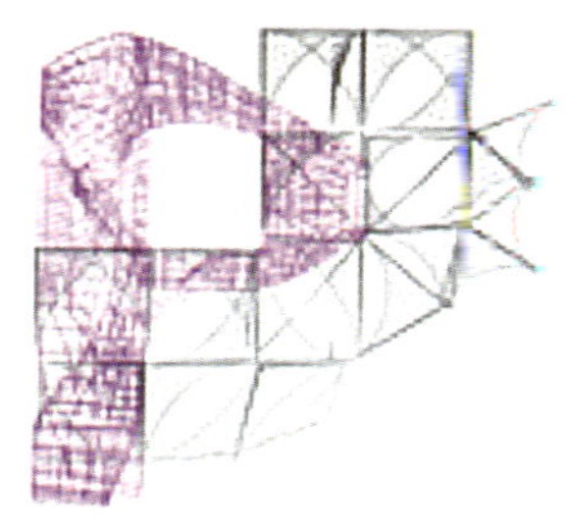

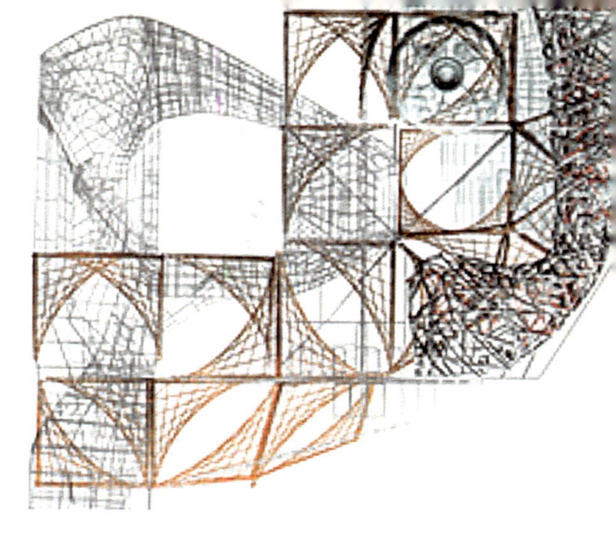

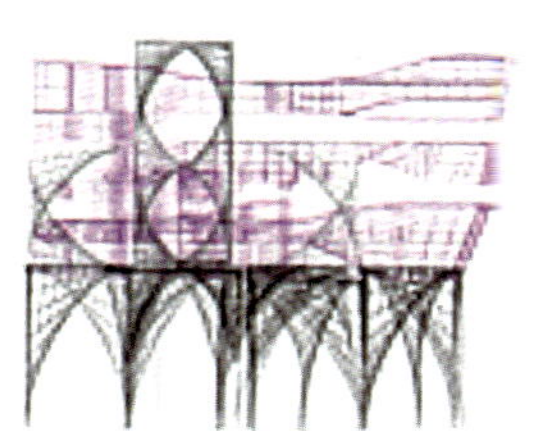

[18]-[01]: infrastructural
[18]-[07]: pedestran flow, car connection
[18]-[09]: structural, pedestrian flow
[18]-[16]: not defined
[18]-]17]: pedestrian flow
[[18]-[19]: infrastructural, pedestrian flow
[18]-[20]: structural, pedestrian flow
[18]-[22]: structural, pedestrian flow
[18]-[23]: structural, pedestrian flow
[18]-[24]: not defined

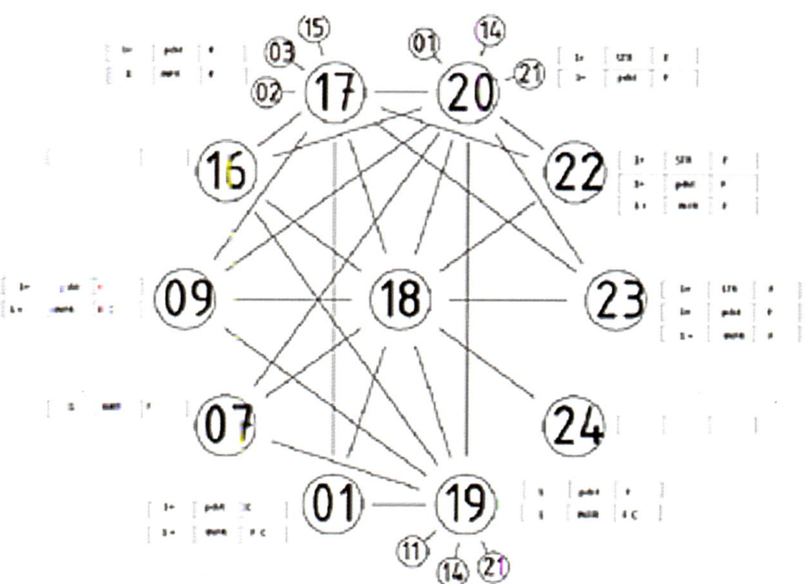

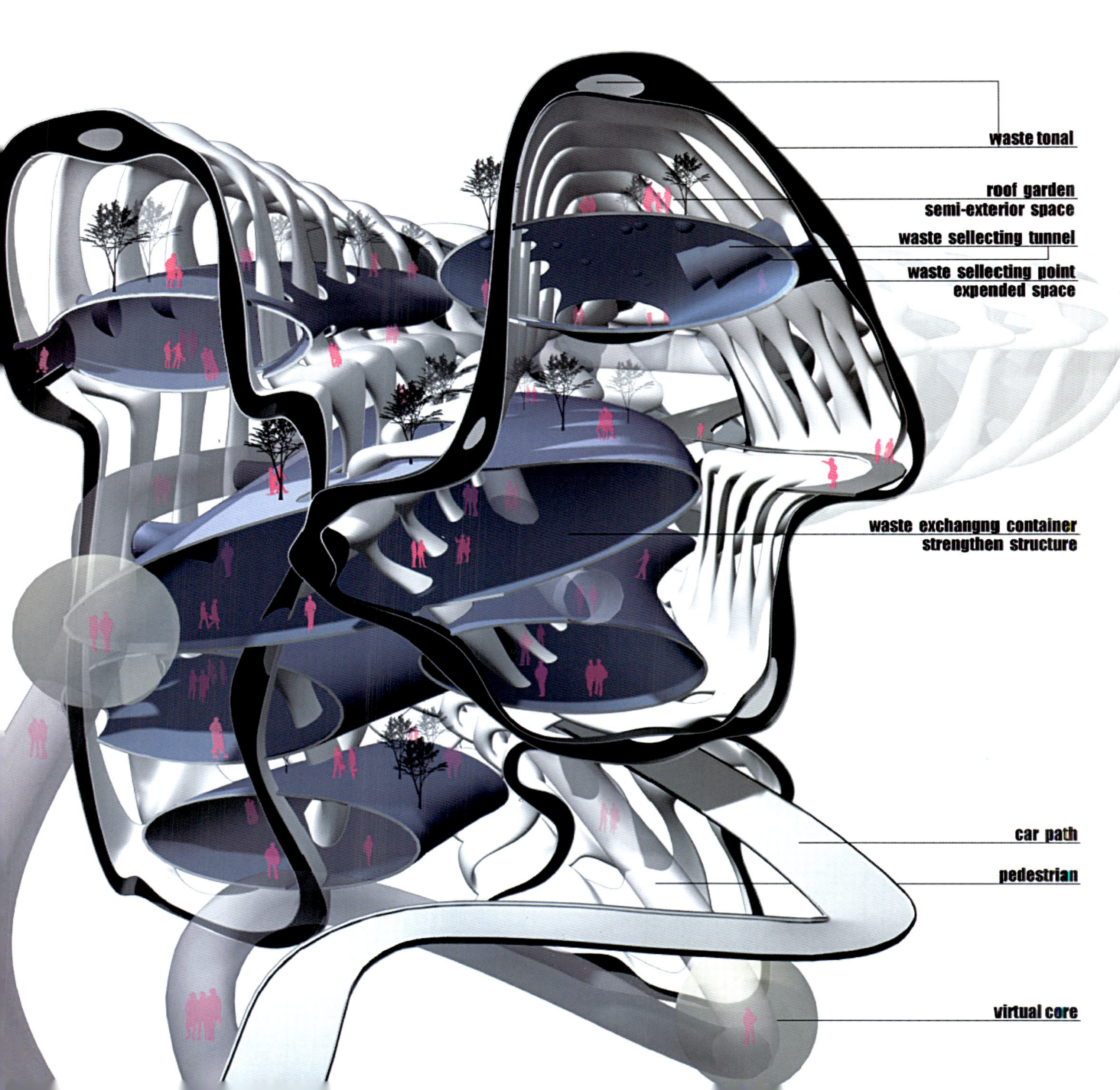
waste tonal
roof garden
semi-exterior space
waste sellecting tunnel
waste sellecting point
expended space
waste exchangng container
strengthen structure
car path
pedestrian
virtual core

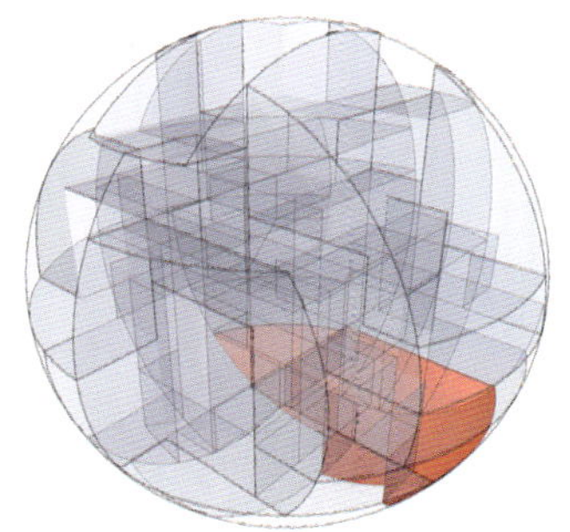

Zone 18 - Stomach
胃
Feng Lin

该建筑像胃一样服务于地处北京751的超合球体。它从周围的地块收集垃圾，其中一部分被来自其他地块的艺术家选为艺术品的材料，剩余的被回收转化为能源供给其他地块。

其他地块的艺术事件将被追踪纪录并转化为抽象的虚拟影像输入到虚拟核心中，而这些虚核根据每两块地块的物理核心在球体中的拓扑关系挤压定位于该地块中。人们在虚核内看到影像时的本能反应，将通过脑电波触发虚核表皮的交互变化。

该地块在球体中的位置以及垃圾传输、转化的功能决定了结构系统的形式。该建筑的表皮系统通过编程控制其生成。只要给定两条曲线，就可以运行程序生成具有细部的建筑表皮。其中的每个构件具有不同的符合力学逻辑的尺寸。建筑的几种主要功能（商业/办公/文化/停车场）服务于共同的主题——垃圾。

This building serves as a stomach of the hybird sphere in Beijing 751. It collects waste from other plots. Some of that waste can be be selected as material for art. The rest gets recycled and transfered into energy.

Art events from the other plots become an input for virtual cores, which are located according to the topology of plot's physical centers. In these virtual cores, people's reaction toward virtual image will activate the skin of the virtual core by the brainwaves of users.

The position of site and waste transfer function together and form the structure of the system. The building has a parametric skin system. So long as there are two curves, a set of non-standard detailed skins can be generated by a running script, according to the logic of force. In this zone, there are several main kinds of functions: commercial, office, culture and parking. All of them relate to the main topic: waste.

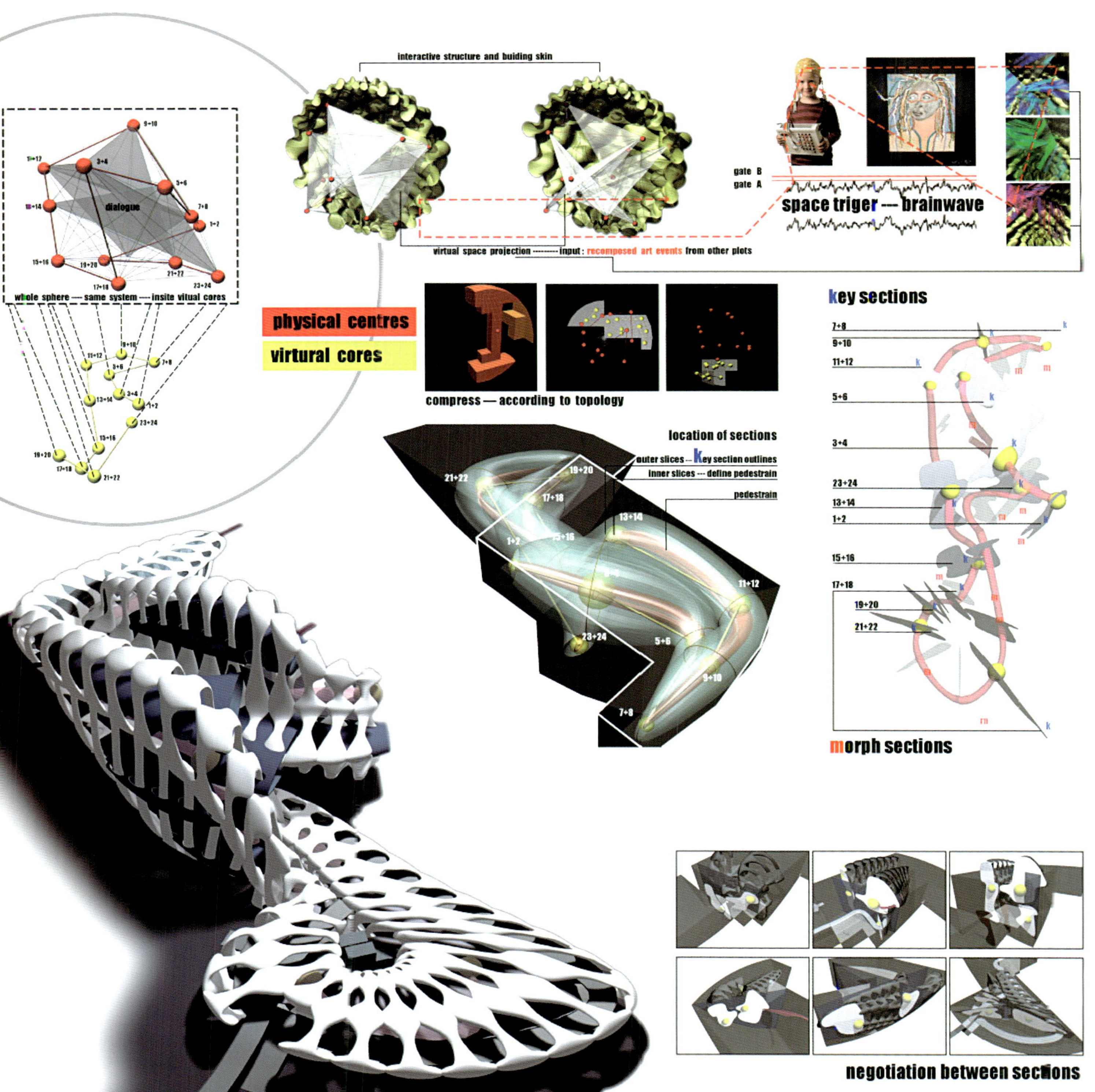
interactive structure and buiding skin
gate B
gate A
space triger --- brainwave
virtual space projection --------- input : recomposed art events from other plots
dialogue
whole sphere --- same system --- insite vitual cores
physical centres
virtural cores
compress — according to topology
key sections
location of sections
outer slices -- Key section outlines
inner slices --- define pedestrain
pedestrain
morph sections
negotiation between sections

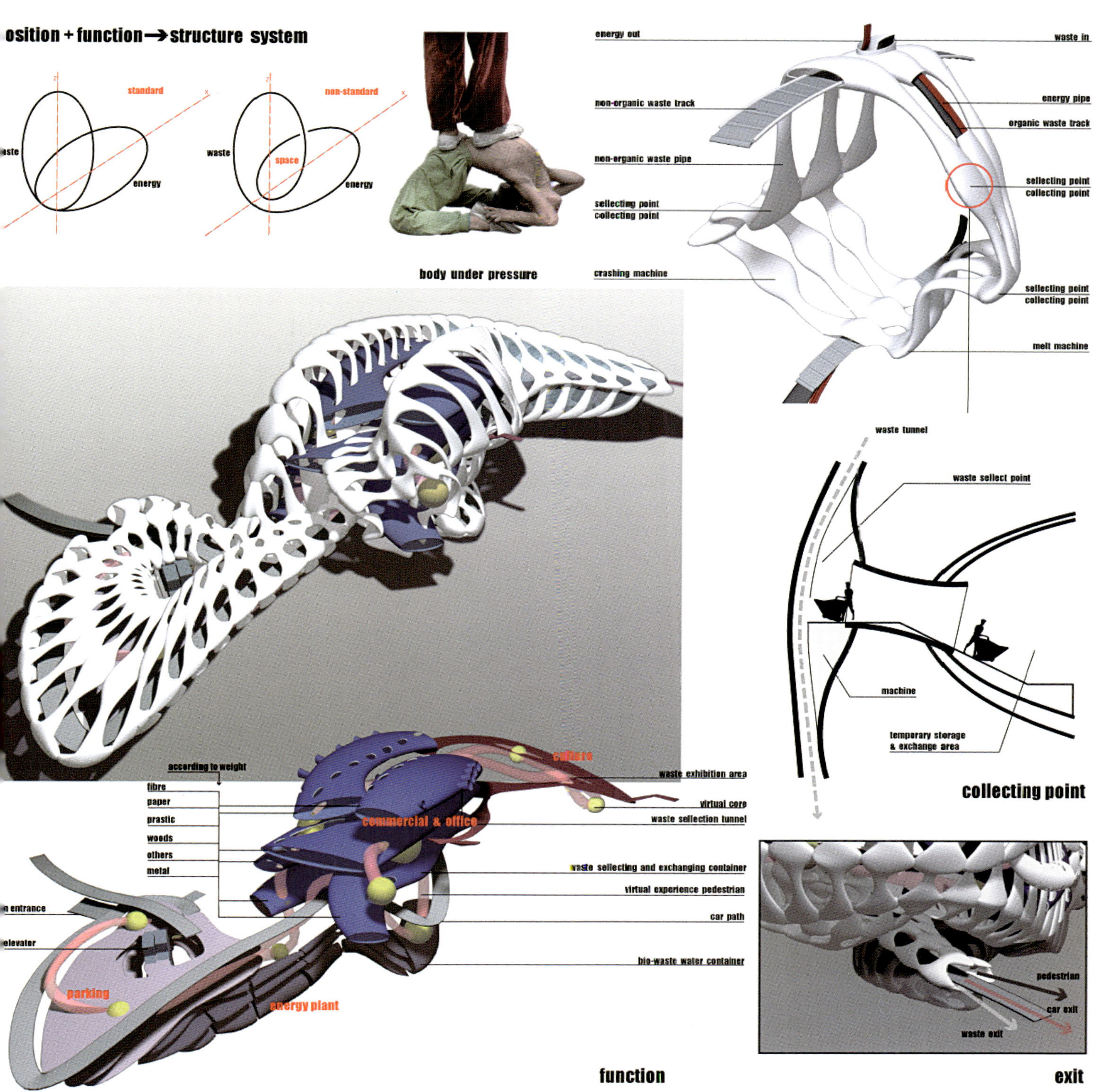
osition + function → structure system
standard
aste
energy
non-standard
waste
space
energy
body under pressure
energy out
waste in
non-organic waste track
energy pipe
organic waste track
non-organic waste pipe
sellecting point
collecting point
sellecting point
collecting point
crashing machine
sellecting point
collecting point
melt machine
waste tunnel
waste sellect point
machine
temporary storage
& exchange area
collecting point
according to weight
fibre
paper
prastic
woods
others
metal
culture
waste exhibition area
virtual core
waste sellection tunnel
commercial & office
waste sellecting and exchanging container
virtual experience pedestrian
car path
bio-waste water container
n entrance
elevator
parking
energy plant
function
pedestrian
car exit
waste exit
exit

semi-exterior or commercial part

connection with neighbors

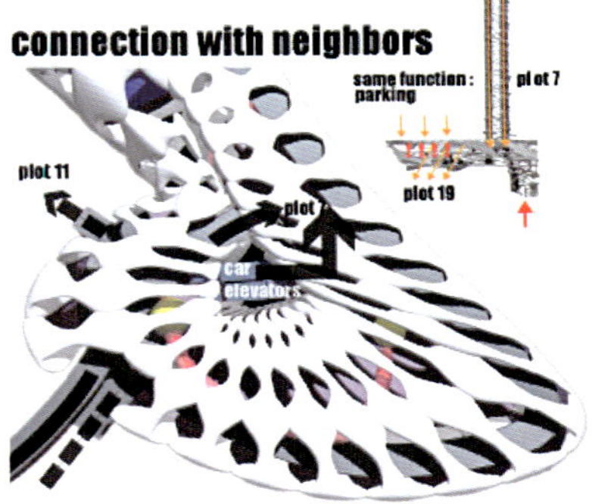

with plot 7 & 11 -- share main entrance

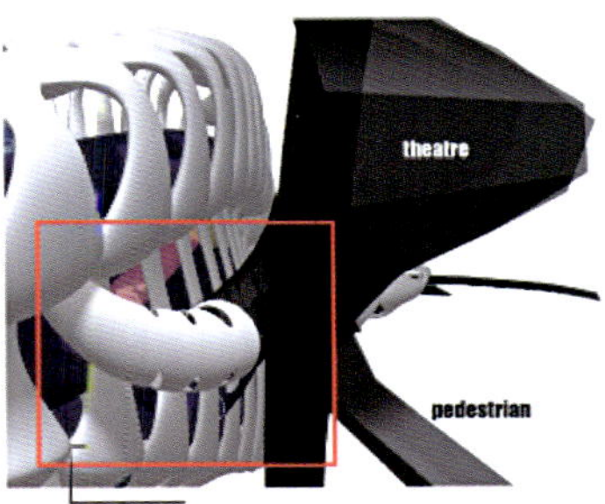

with plot 15 -- share pedestrian and culture

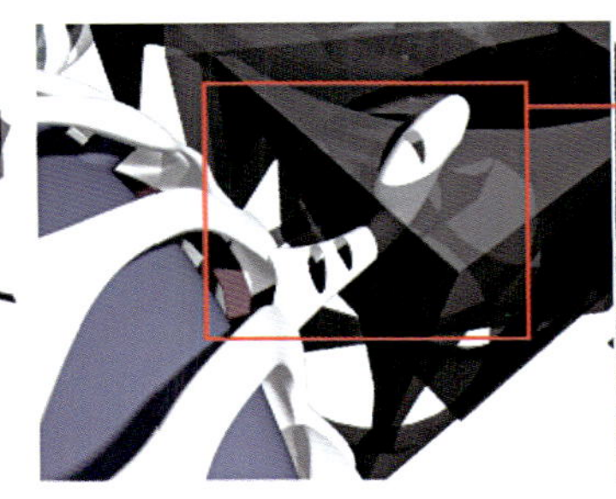

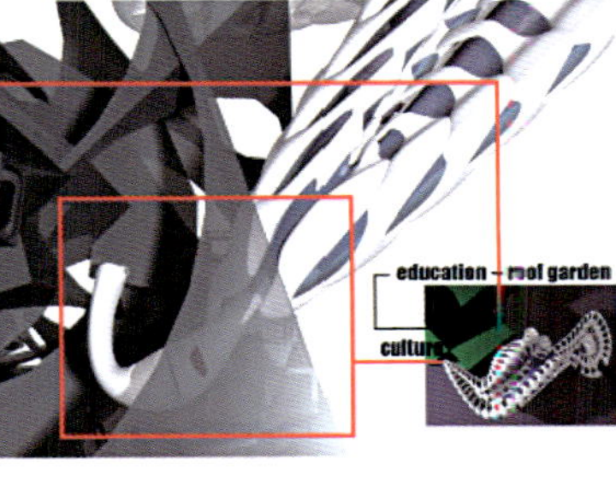

with plot 18

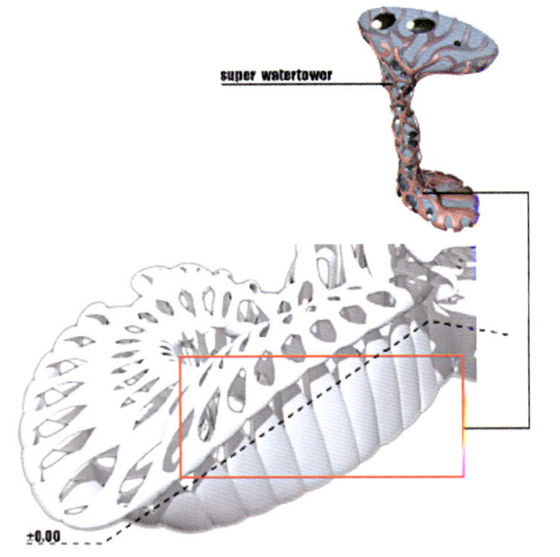

with plot 1

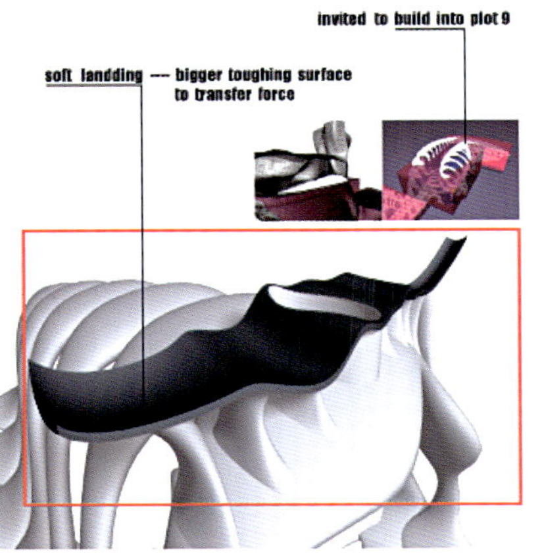

with plot 9

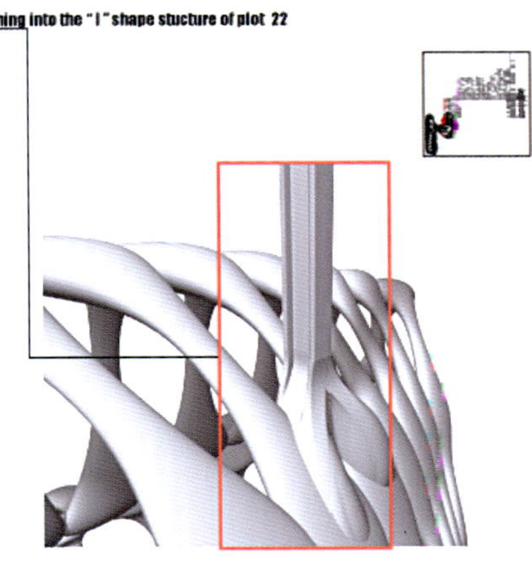

with plot 22

interior of culture part

rotatable openning system

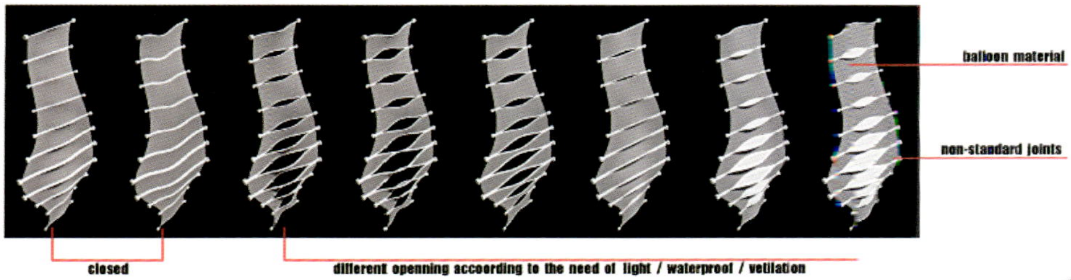

balloon material

non-standard joints

closed

different openning accoording to the need of light / waterproof / vetilation

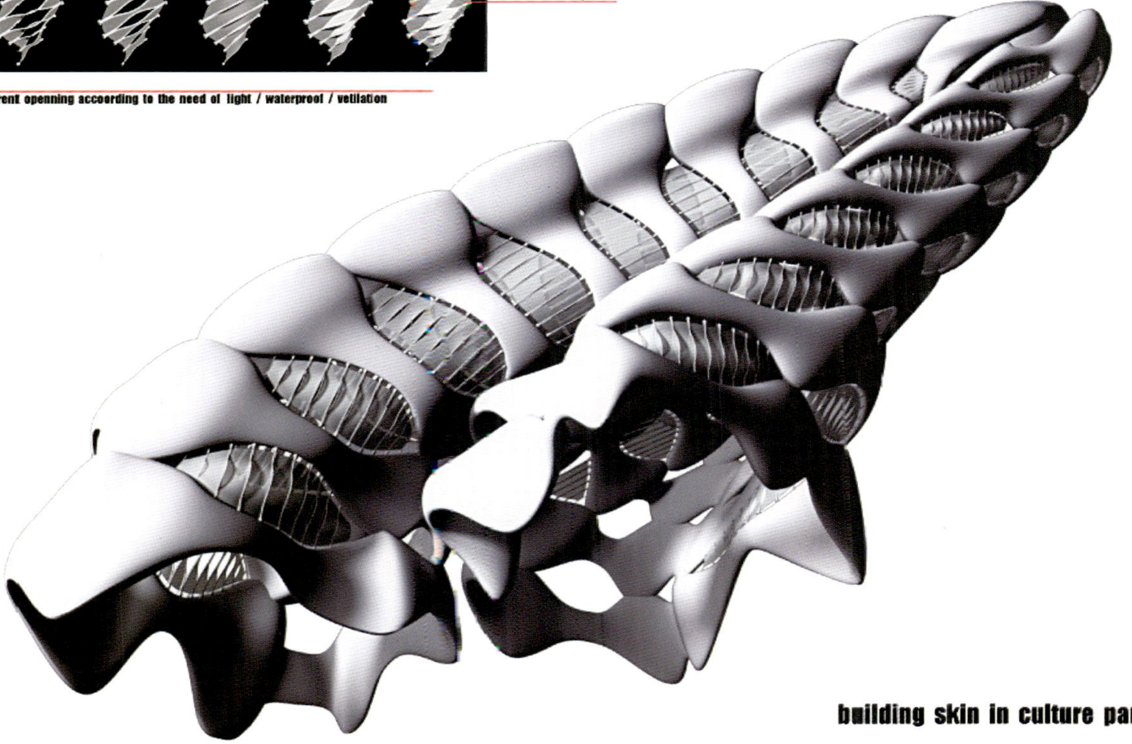

building skin in culture part

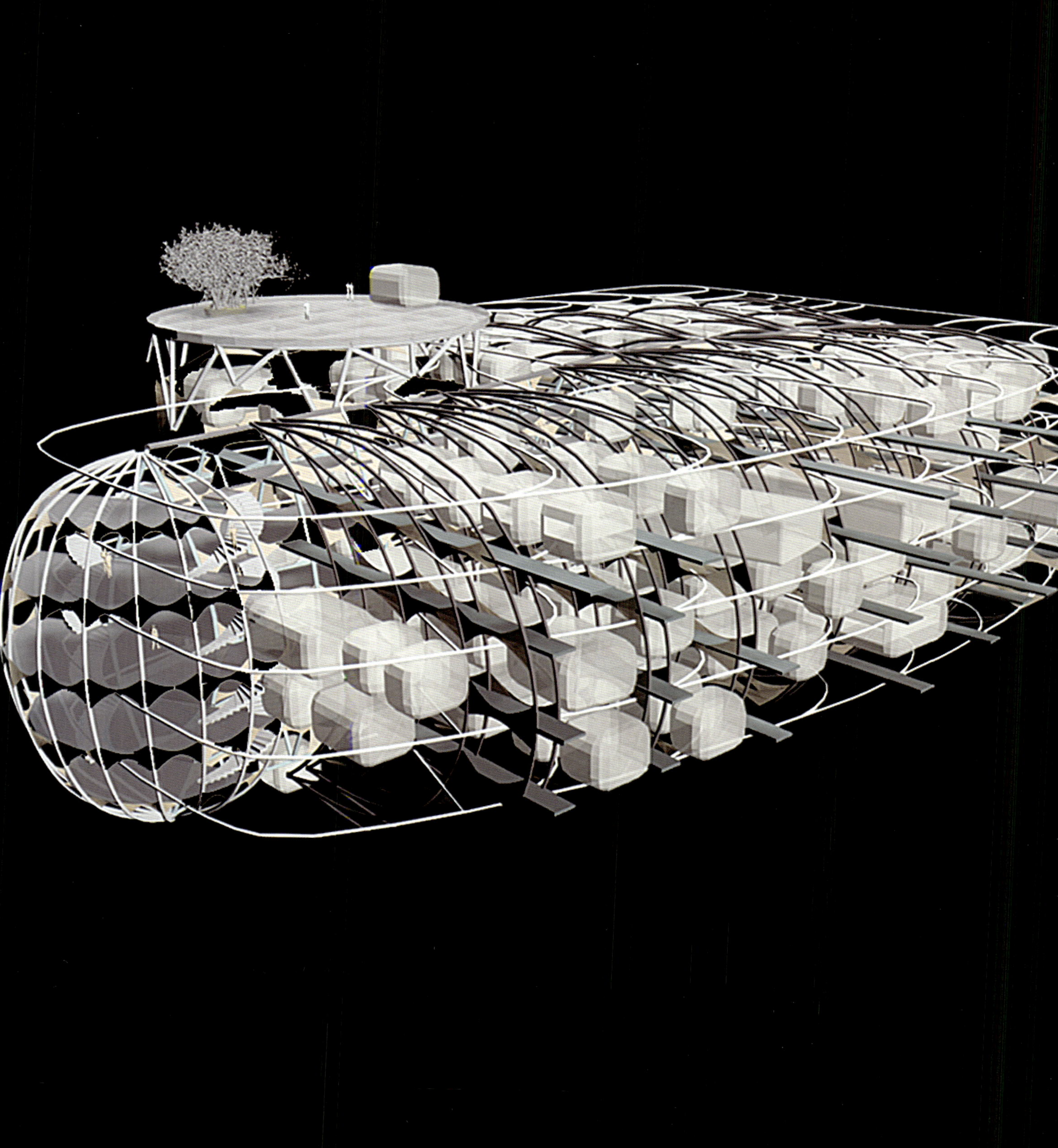

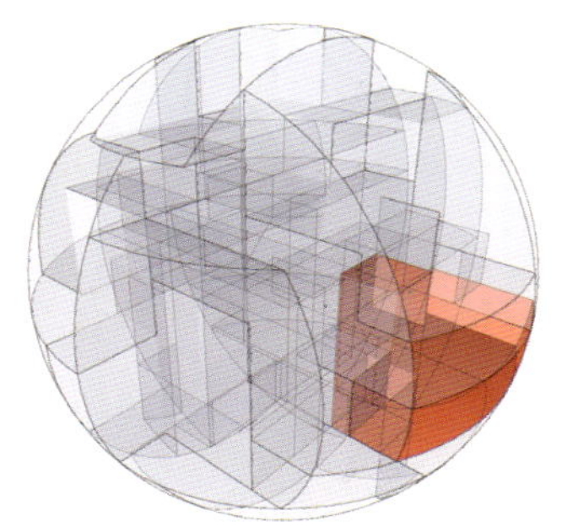

Zone 19 - Mechanical Spaceship
机械太空船
Julien Westerfeld

超球体的第19区位于地面以上70m，是一个“机械太空船”，伸展开其所有的钢铁臂膀和超过218000m³的设施，包括163000m³的开放空间和54500m³的建成空间。功能包括4000m³的休闲空间，诸如电影院、咖啡馆和一个体育中心；2500m³的艺术教育功能，诸如艺术学校，包括音乐、舞蹈、剧院、绘画和雕塑；5000m³的居住小屋，包括露台和其他休憩空间；2600m³的商业区域，如百货商店和小商店；15600m³的文化空间，包括工作室、艺术家休息室和展览厅；最后是至少16000m³的办公建筑，主要供艺术家和该区域的艺术管理部门使用。每个小屋都悬挂在一个整体钢结构上，该结构允许交通活动，这决定于整个综合体的特定需要。因此，不同的时间参数可与不同活动的时间表相联系，比如展览、事件、在超区域居住或工作的人员数量，或者相邻区域对于小屋的需要。所有这些通过艺术家和他们的行动和分配，定义了小屋的“艺术与工艺”产品。同时在接下来整个综合体的演进中，产品、作品、交通、分配和交互的不同参数之间的相互作用，将参照用户的临时需要，定义各类不同的空间。

Located 70m above the ground, zone 19 of the hyper-sphere, a sort of a “mechanical spaceship”, spreads out all its steel arms and facilities over an area of 218 000m³, including 163 000m³ of open space and 54 500m³ of built space. The program includes 4000m³ of leisure spaces such as a cinema, cafes and a sport centre; 2500m³ for art education such as school of art, including music, dance, theater, drawing and sculpture; 5000m³ of residential cabins, including terraces and other relaxation spaces; 2600m³ of commercial area such as grocery stores and basic shops; 15 600m³ of cultural spaces containing studios, sleeping places for artists and exhibition halls; and last but not least 16 000m³ for office buildings which are mainly used by artists and art administrations of this district. Each cabin hangs on a global steel structure which permits traffic depending on the specific needs of the whole complex. For this different time parameters can be related to time schedules of different activities such as exhibitions, events, amount of people working or living in the hyper-zone or the zone neighbors' need for cabins. All this defines the amount of “art and craft' production of cabins by artists and their implementation and division. Then following in real time the evolution of the whole complex, the interaction between different parameters of production, creation, transportation, division and interactivity will define different kinds of spaces referencing to temporary needs of users.

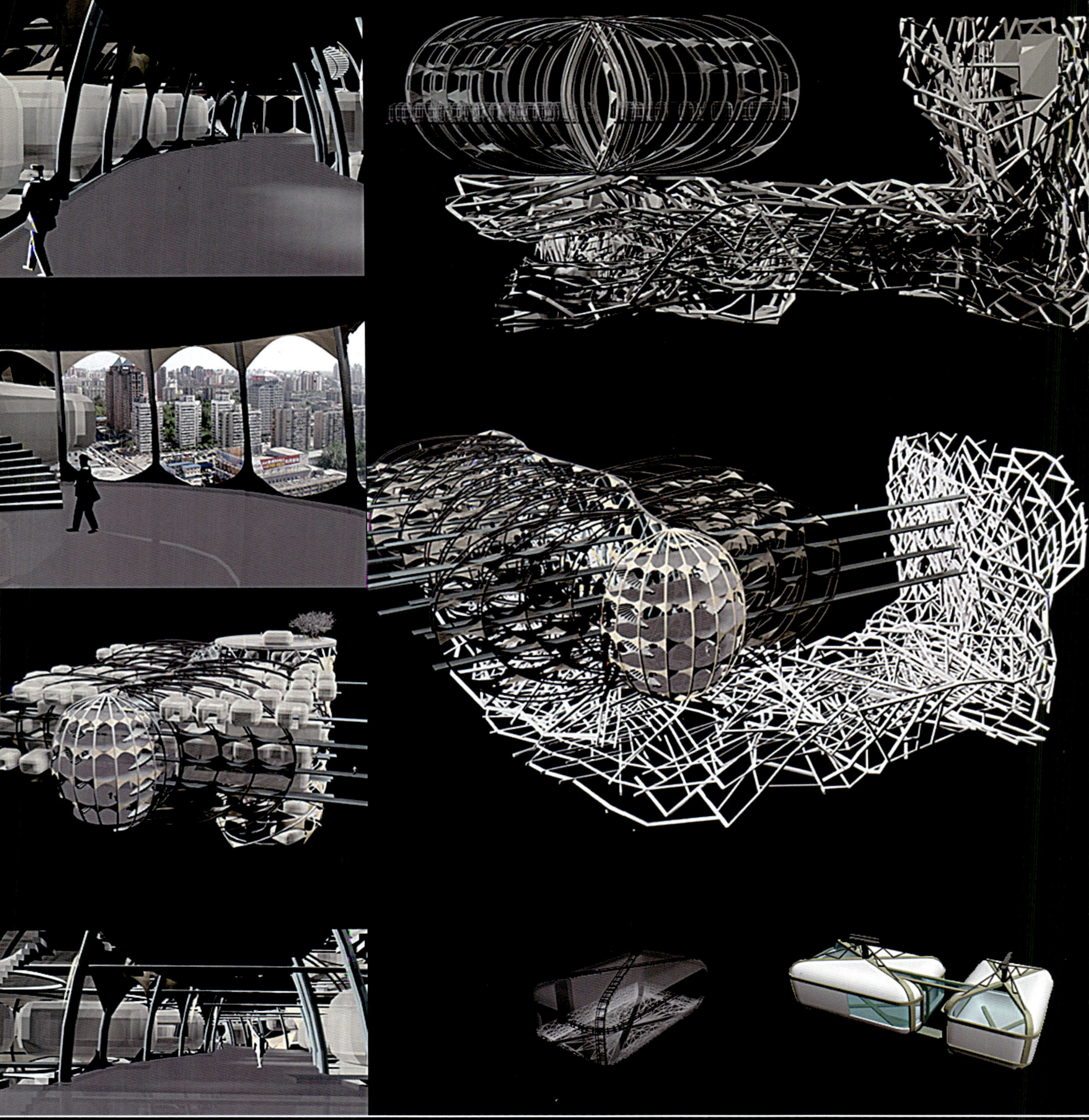

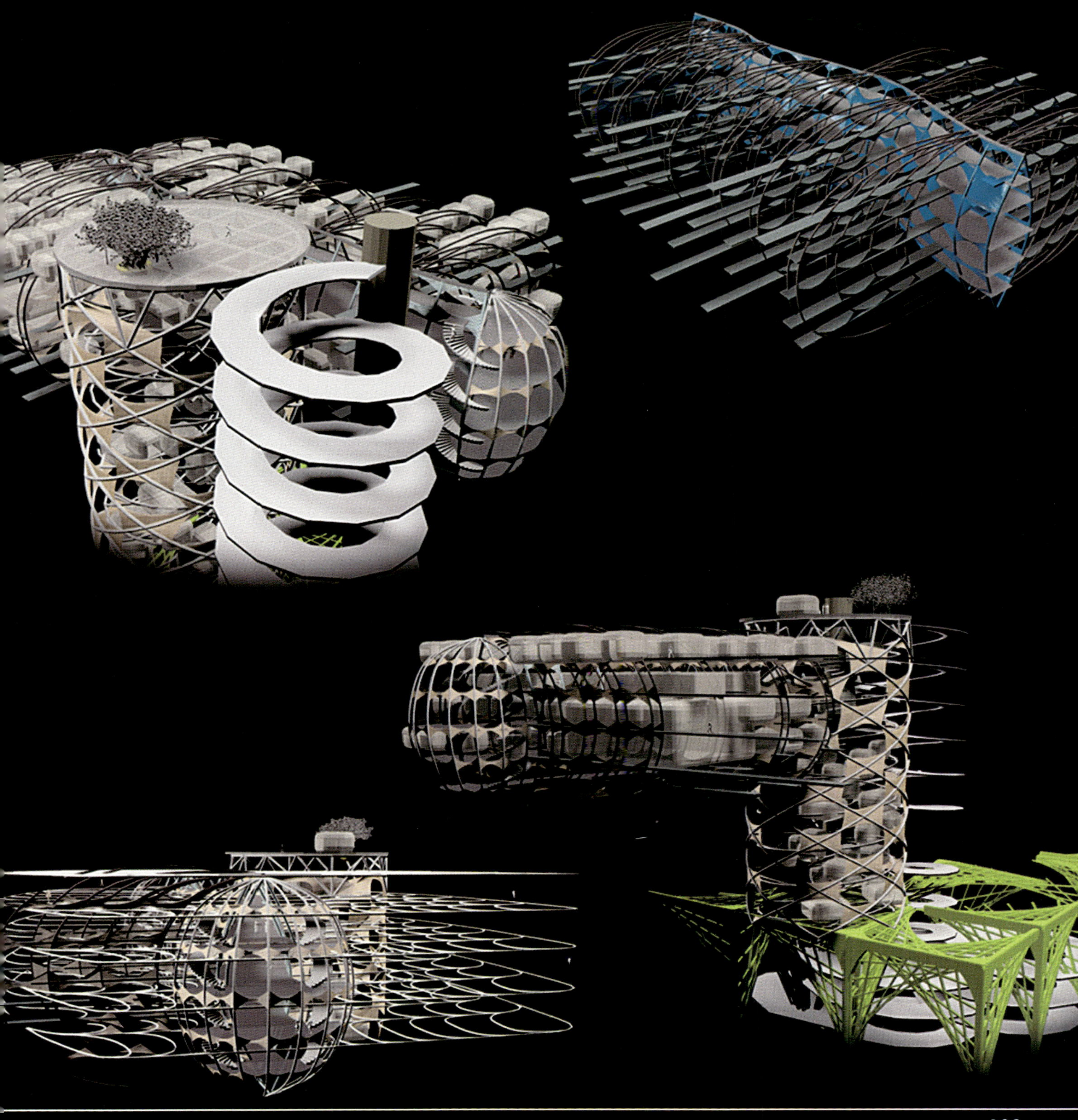

inside view
cfc

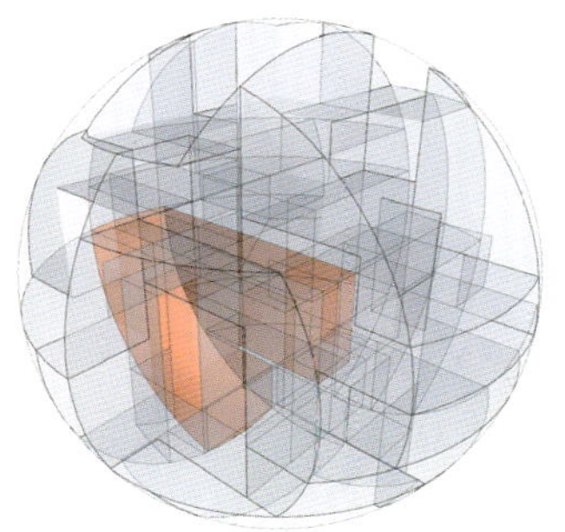

Zone 20 - Culture Flying Circus
飞行文化马戏团

Lukas Mahlknecht

在北京真正的高密度区域，生活品质确实很差：空气污染、灰尘、燥热、噪声，没有空间静下来好好放松。

与此相对立的是，分散在这个搏动着的城市中的那些巨大公园。很多人在寻找宁静的空间，以练习功夫、演奏音乐或者午餐休息。

但这些文化绿洲固定在原地，不能对不同地方的空间需要作出适时反应。

飞行文化马戏团能够飞行，哪里需要展览、音乐会、电影院、娱乐、灵活空间等，都可以作出反应。它降落在街道、广场、摩天楼之间、水面上或者干脆就是飞行，以其源自三维墨比乌斯带的复杂造型吸引着人们的关注。

The quality of life is in the really dense zones of Beijing really small: polluted air, dirt, hectic, noise, no space to come down and relax.

The opposite are the huge parks scattered over the pulsating city. Many people are finding there a quiet space, doing Kung-Fu, playing music or having their lunch brakes there.

But this green culture oases are fixed in place and can not react to the punctual need of space in different locations.

The culture flying circus has the capability to fly and is able to interact where a demand is given for exhibitions, concerts, cinema, relaxation, very flexible spaces etc. Landing on streets, squares, between skyscrapers, on water or just flying, it attracts attention of people with its complex shape inspired by a threedimensional Möbius-Band.

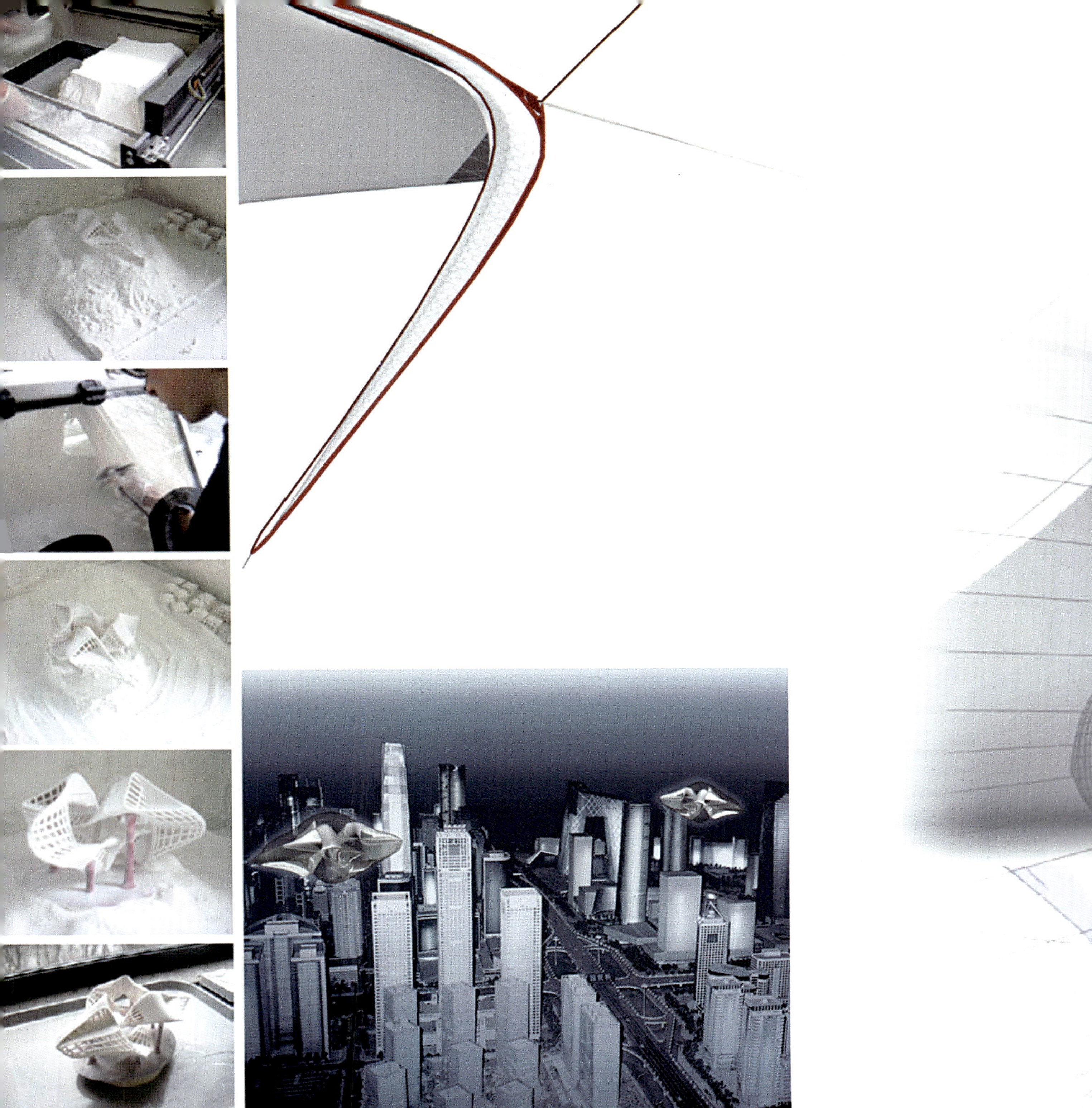

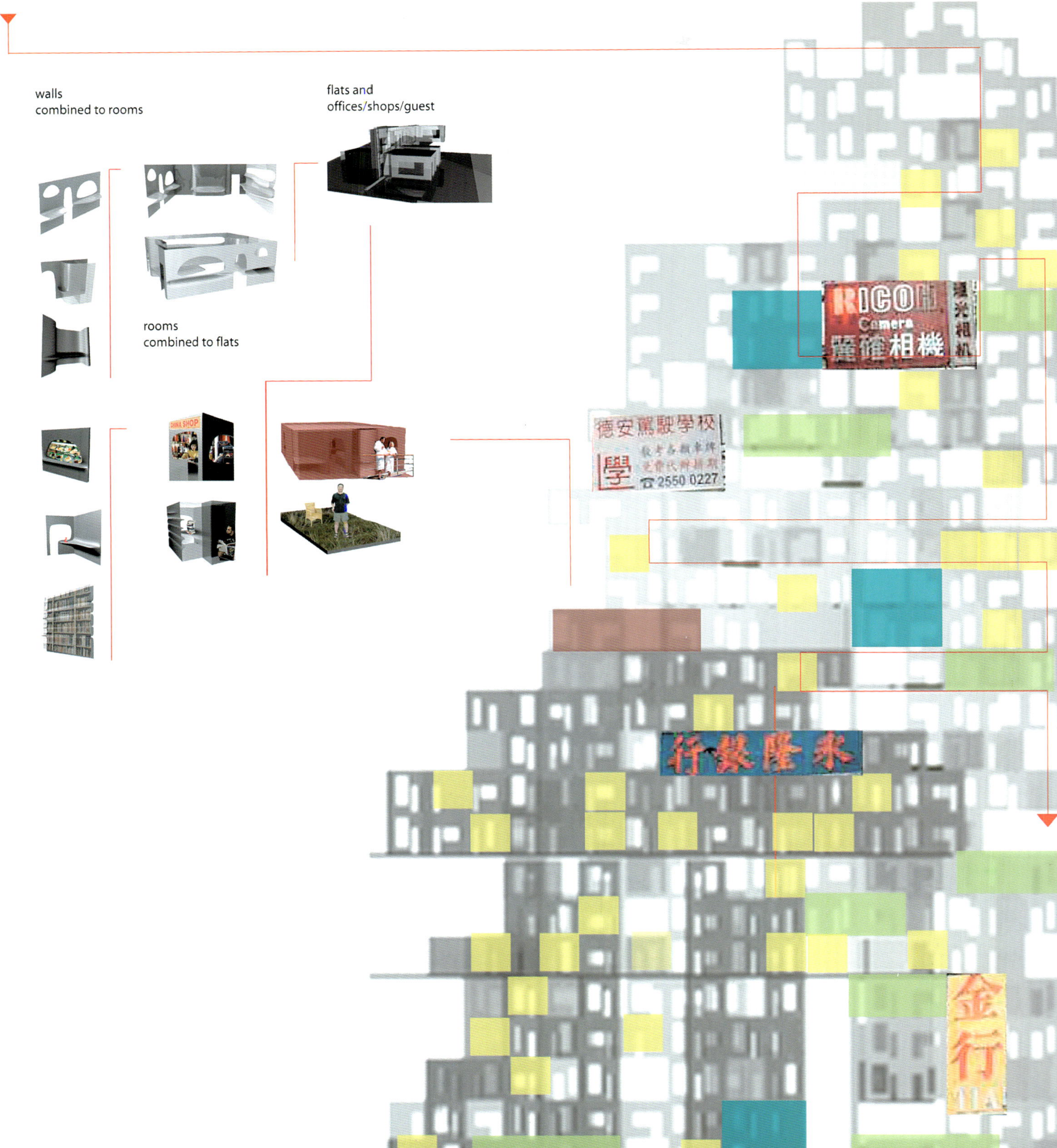
walls
combined to rooms
flats and
offices/shops/guest
rooms
combined to flats
Camera
德安駕駛學校
2550 0227

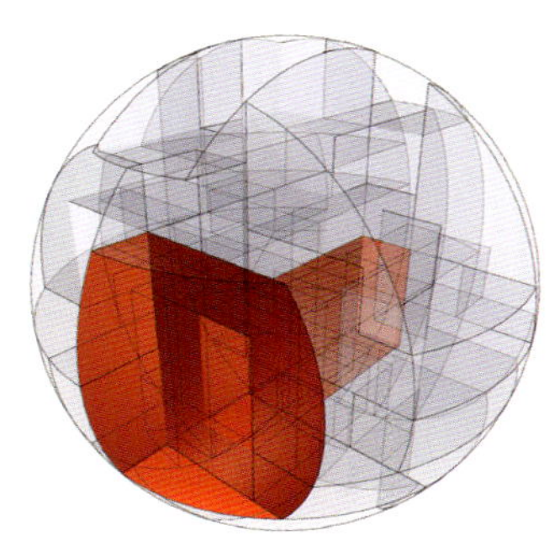

Zone 21 - Out of Town
城外

Astrid Mitterlehner

不同功能的小型元素被安置在球体中，就像一个贮藏室。他们可连接起来形成一个工作－居住－文化－环境。他们可散布在整个城市中。他们可根据需要改变自身位置。

在球体中，这些元素吸附于一个1m厚的透明墙体，该墙体就像一个充电站，因为它为这些单元供应来自于太阳和光电电池的热水和能量。使用者能够看到透明的太阳和光电电池表皮背后的所有电缆、管道、柱子和线路。人们应当知晓其能量的使用。

Small elements with different functions are settled in the sphere, like in a storage. They can be connected to create a working- living- cultural- environment. They can spread out all over the city. They have the possibility to change their position by needs.

In the sphere these elements are attached to a one meter thick transparent wall which works like a charging station, because it supplies the units with warm water and energy, which will be created by solar and pv cells. The user will be able to see all the cables, tubes, batteries, columns and tracks behind the transparent solar and pv cell skin. People should be aware of their usage of energy.

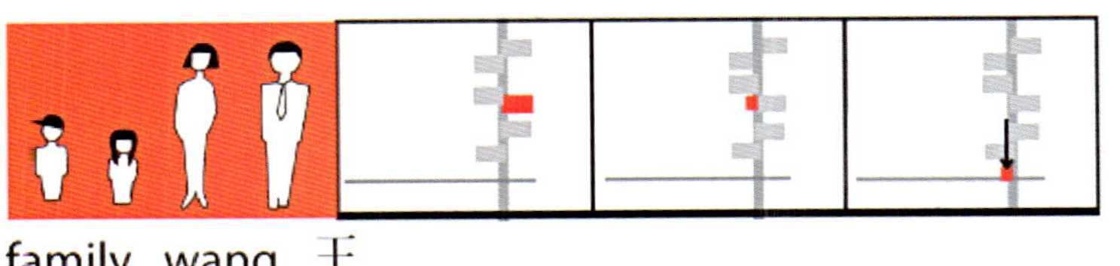

family wang 王

out of town

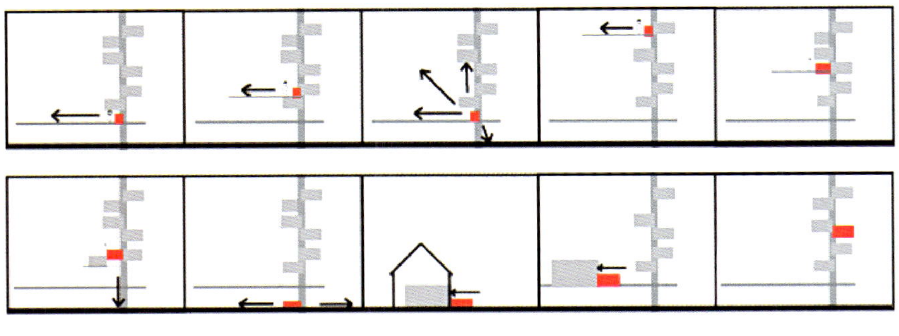

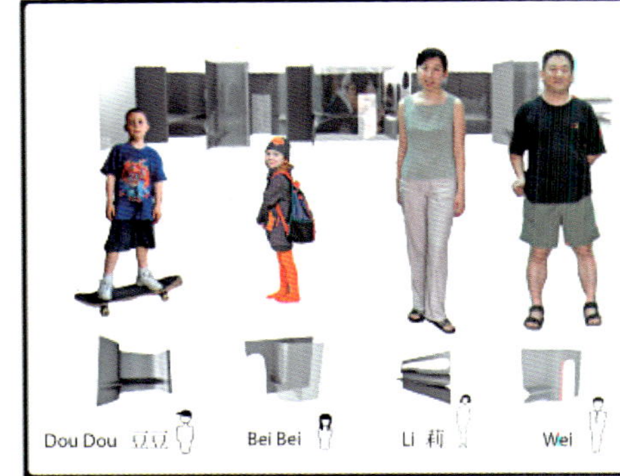

family wang lives in beijing

every morning they are getting up facing the south of the city

dou dou and bei bei leave together every day

in the big public square they are heading off in different ways

bei bei is going to sema's school

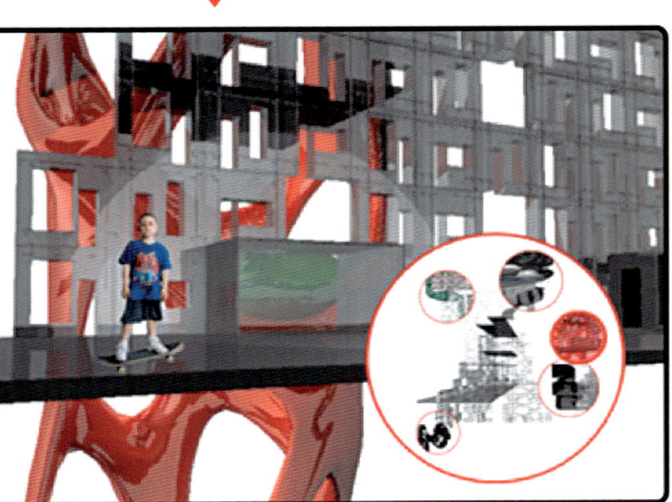

dou dou should go to school too but instead he explores the neigbourhood

public space

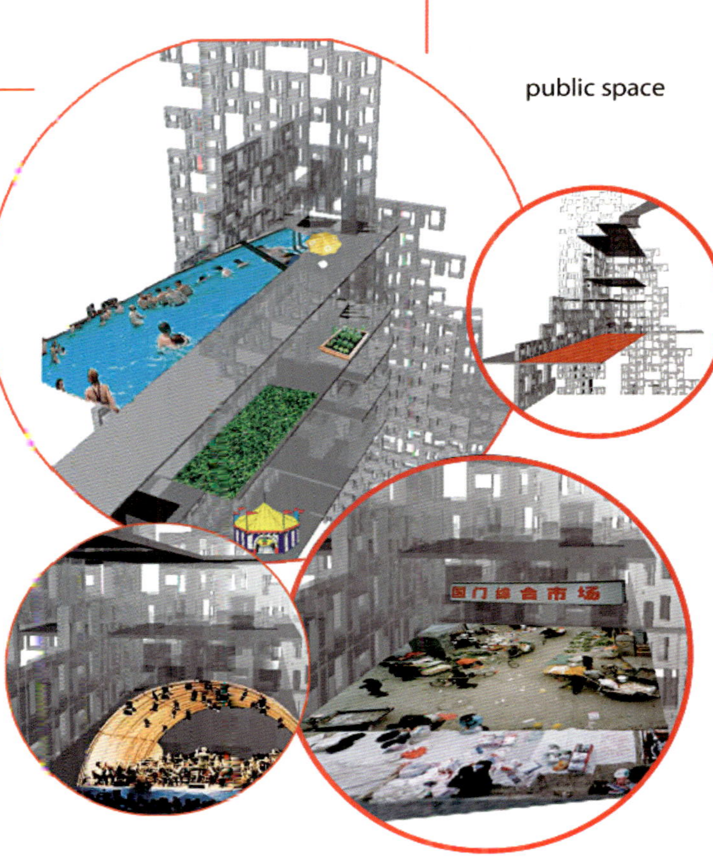

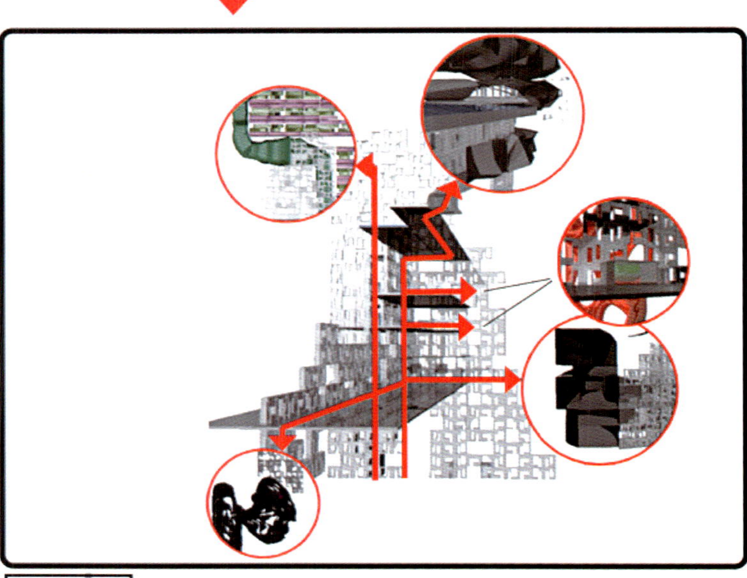

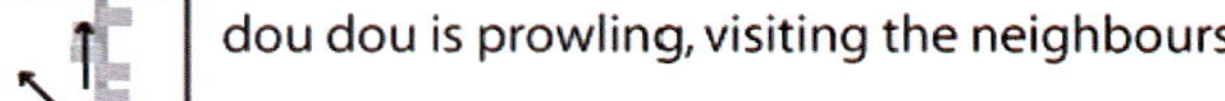

dou dou is prowling, visiting the neighbours

hurrying to work? li is there already
the shop is right next to the living studio

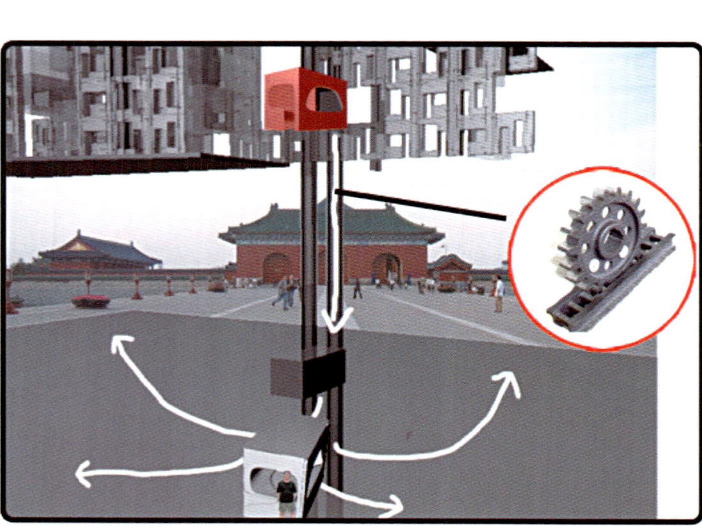

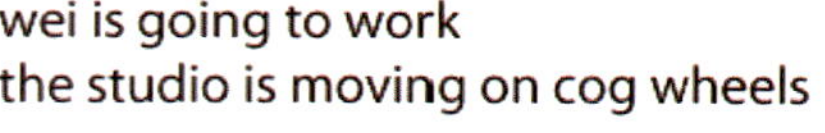

wei is going to work
the studio is moving on cog wheels

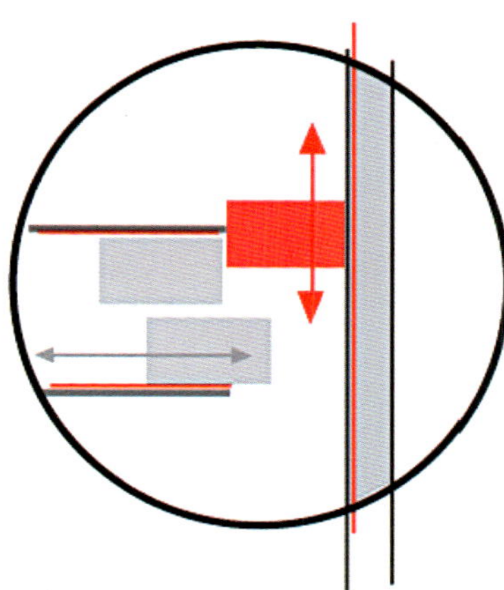

when a studio is moving vertically
they others can move horizontally backwards on
the tracks

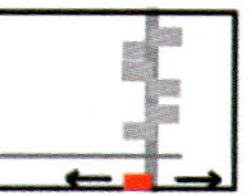

wei is on his way to work
in the streets of beijing the studio is moving
on wheels

when wei is working in the centre
the studio docks to the common room
in an old house

when wei is working with others in the sphere the
studio docks to a common room on a public square

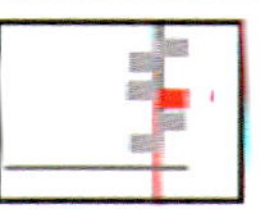

the whole family is back at home in the sphere
the studio is pluged in to charge
and tommorrow they might stay in the centre....

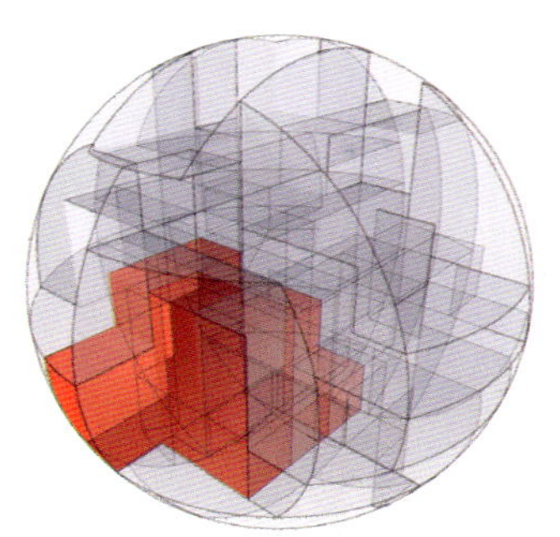

Zone 22 - Made in Space
空间制造

Tobias Schimme

空间制造设计于室外，也就是外部运动。它将球体内没有直接联系的相邻地块连接起来。所以它围绕水塔建立了一个条带，为几个设计提供了通往停车塔的捷径。内部的多重功能相互叠加和交互：电子音乐工作室与俱乐部空间相联，旅馆可从一个北京的观景台出入，该观景台也是商业空间的起始点。停车空间位于建筑底部并可通往所有功能。

运动沿斜向和垂直方向（3个节点），以及水平方向（节点间的桥梁）组织。

在节点中增加了改变方向和功能的可能性，楼梯弯曲、向下并限定空间，透明电梯描述了垂直运动，虽然他们在桥上时是一个方向。

Made in Space is designed out of inner as well as outer movements. It connects neighbors that have no direct contact inside the sphere. So it creates a band around the watertower, and gives shortcuts for several designes to the parking tower. Inside multiple programs overlap and interact with each other - studios for electronic music are linked to clubbing spaces, a hotel is accesible from a view platform over Beijing, the platform is the starting point for commercial space. The parking space lies in the bottom of the building and gives access to all functions.

The movement is organized in diagonal and vertical directions (3 nodes) and horizontal (bridges inbetween the nodes).

The possibility to change directions and functions increases in the nodes, stairs bend up - and downwards and define space, transparent elevators describe vertical movements, while they are one directional in the bridges.

CONCEPT DIAGRAM
PRIVATE
SEMI PUBLIC/PRIVATE
PUBLIC
PARKING
electronic music
STUDIOS, LOUNGE, REHEARSAL ROOMS, OFFICE
club
digital art
ATELIERS, LOUNGE, AUDITORIUM, OFFICE
exhibition
shopping
bar/ restaurant
hotel
ROOMS (SHORT&LONG STAY
theather
dance
STUDIOS, CHANGING ROOMS, LOUNGE

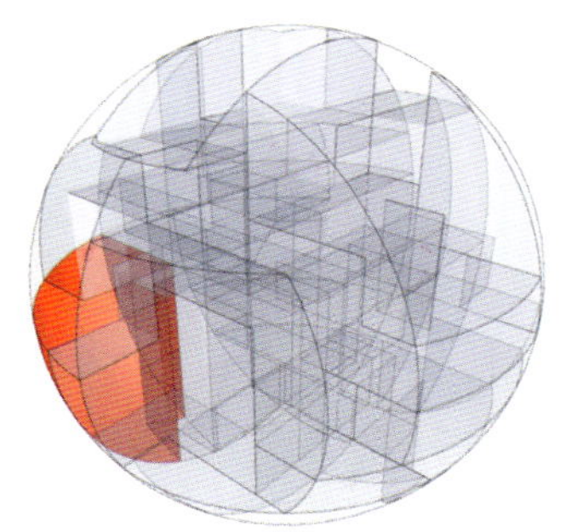

Zone 23 - Intermediator
中间体

Theresa Schütz

我将公共空间作为一个讨论的平台：一个你可以作为人工景观使用其内外的构筑物。通过吸纳相邻地块的需要并将其意图通过我的地块来传输，而将我的设计作为中间体来运作。我用参数设计来将功能输入－输出的图表转译为形式，就像一个倒转的楼梯。

1．孢子作为模板上的一个层在铸造时与另一个丰富的海面材料层一起直接作用于混凝土，在水泥构件上留下孔洞，这些孔洞可作为景天类植物的生长根基。

2．玻璃开口作为直接顶光，而当日光碰到与混凝土表面相连的光纤玻璃时，光线就通过该材料进一步传入建筑的内核，作为间接光源。

3．支撑结构＝生长根基＝光线载体

4．人工景观提供了步入其内部或外部的可能性，减弱了内外之间的障碍。

I provide public space as a platform for discussion – a structure, which you can use inside and outside as an artificial landscape.
By absorbing the needs of my neighbours and transmitting their offers through my plot my design shall work as intermediate.
I used parametric design to translate a functional input-output diagram to a form which works like an inverted staircase.

1. spores as a layer on the formworks directly onto the concrete during casting in cooperation with another layer of fertile sponge material leaves holes in the cement structure as a growing basis for sedum plants.
2. glass openings serve as direct overhead light while sunlight hitting the fibre glass connected concrete surfaces is carried through the material further into the core of the building and serves as indirect light source.
3. supporting structure = growing basis = carrier of light
4. artificial landscape gives the possibility to walk inside and on the outside of the building and loosens the barriers between inside and outside.

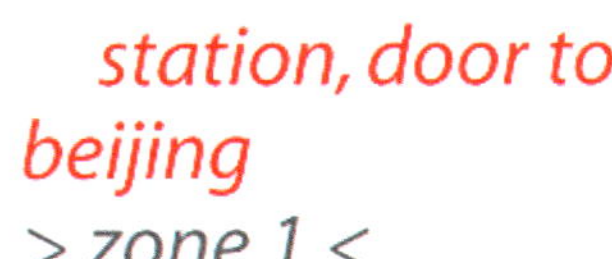

station, door to beijing
> zone 1 <

sports cente
> zone 2 <

vertical structural support
> zone 3 <

drive-in-cinema
> zone 5 <

service area
> zone 17 <

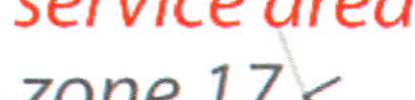

open-air udience hall
> zone 20 <

short-stay offices
> zone 22 <

artificial landscape
> beijing < <

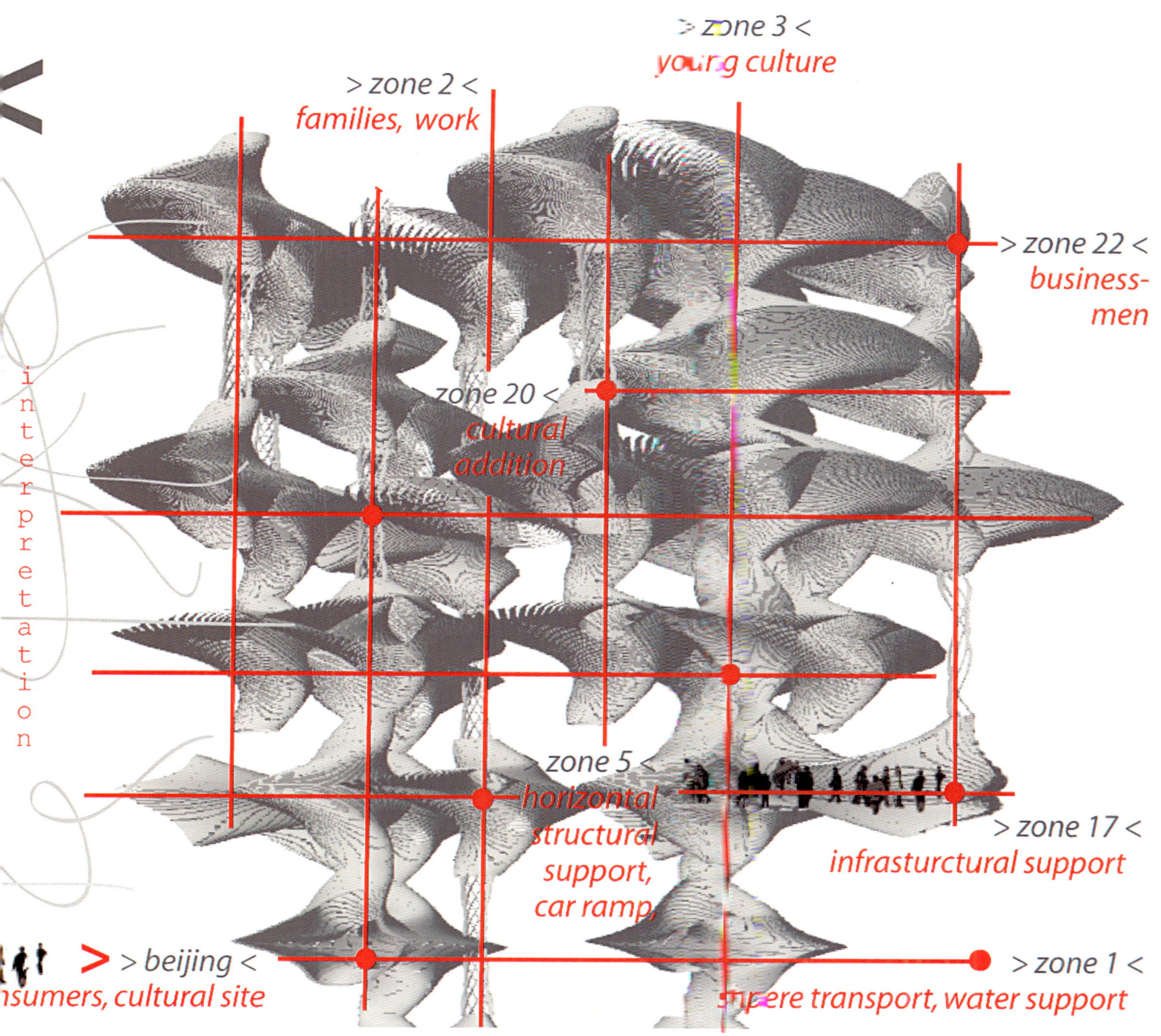

the functional output of my neighbours is the functional input for the intermediator

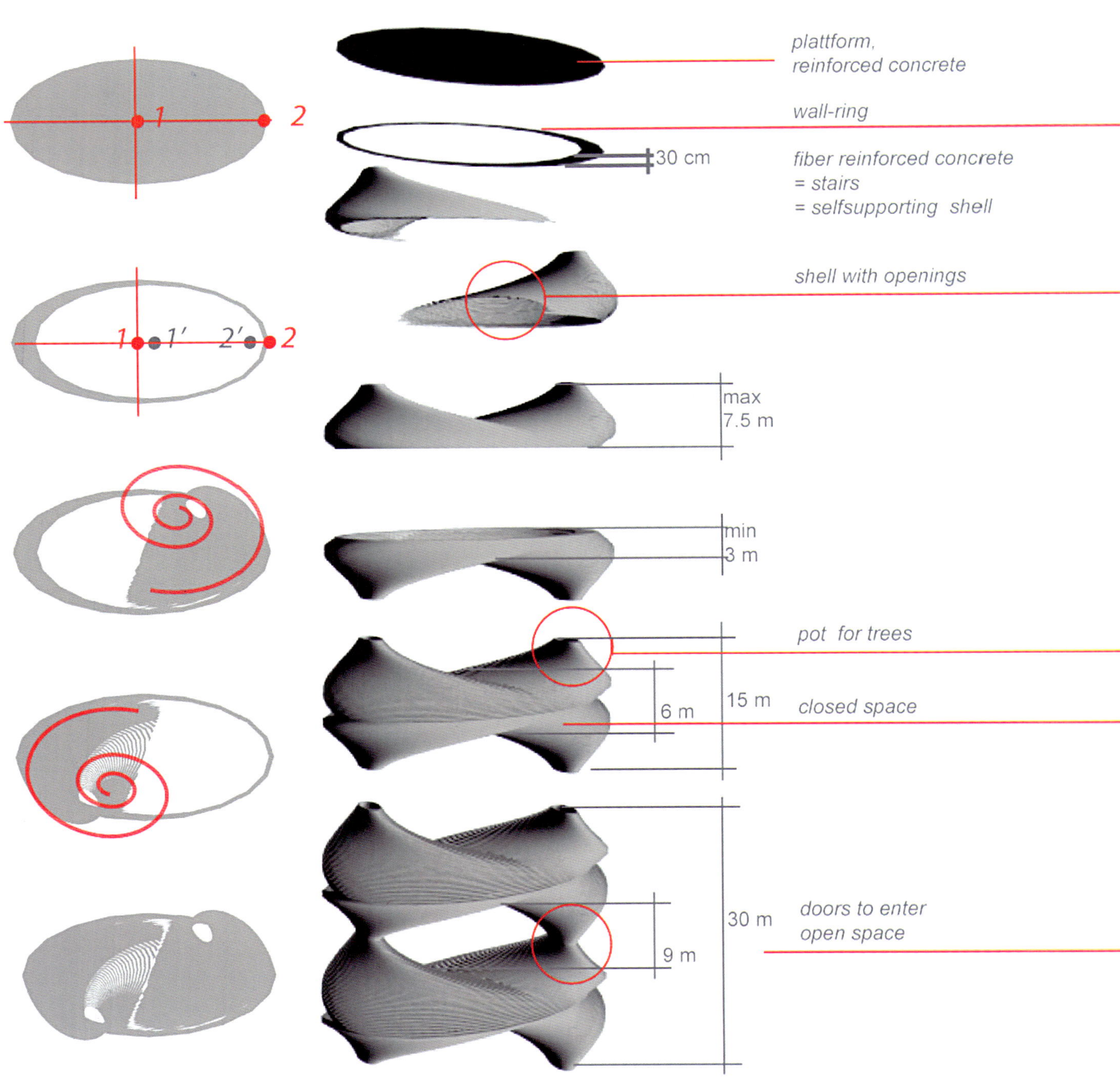
1
2
plattform,
reinforced concrete
wall-ring
30 cm
fiber reinforced concrete
= stairs
= selfsupporting shell
1
1′
2′
2
shell with openings
max
7.5 m
min
3 m
pot for trees
6 m
15 m
closed space
30 m
9 m
doors to enter
open space

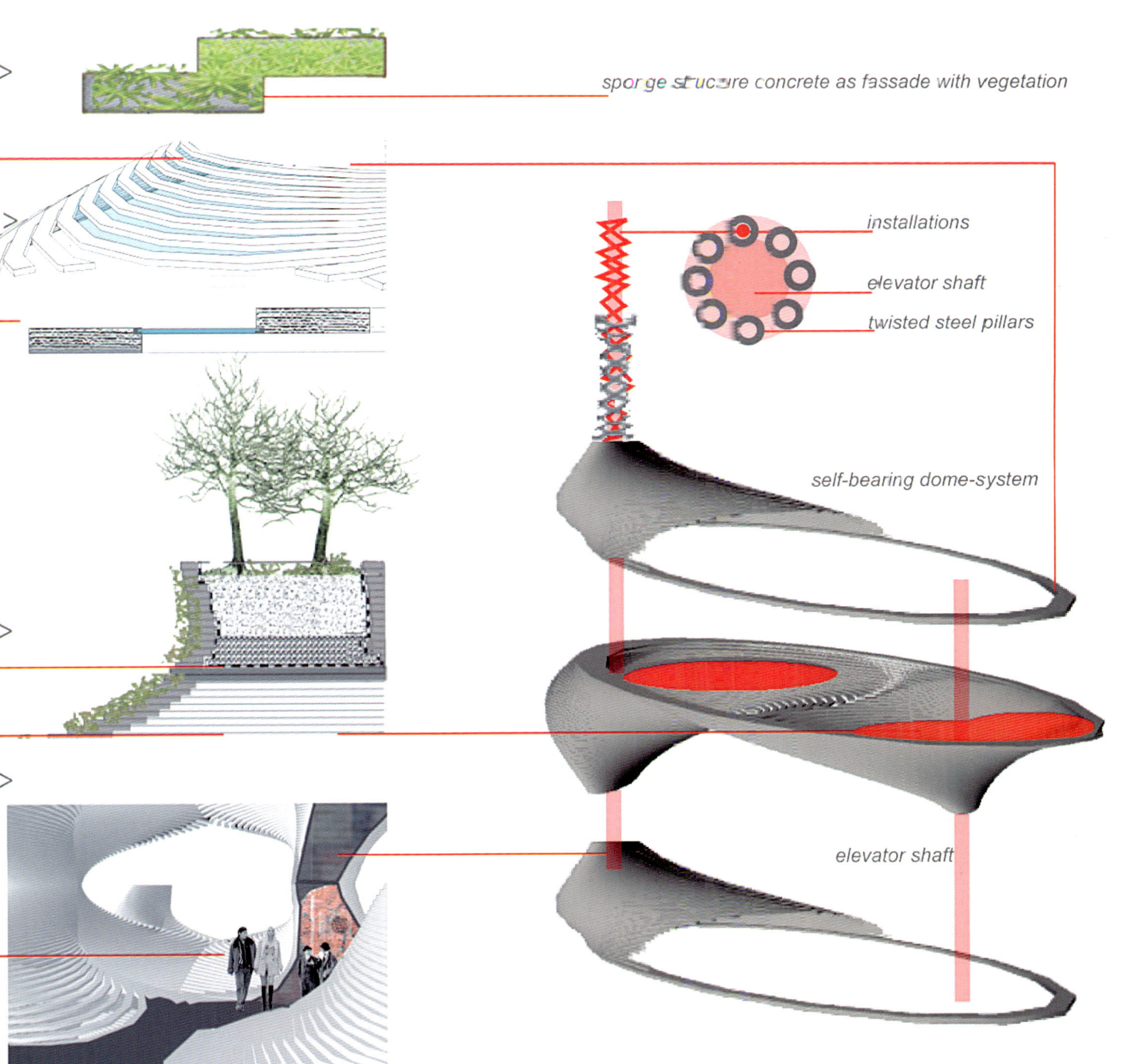

sponge structure concrete as fassade with vegetation
installations
elevator shaft
twisted steel pillars
self-bearing dome-system
elevator shaft

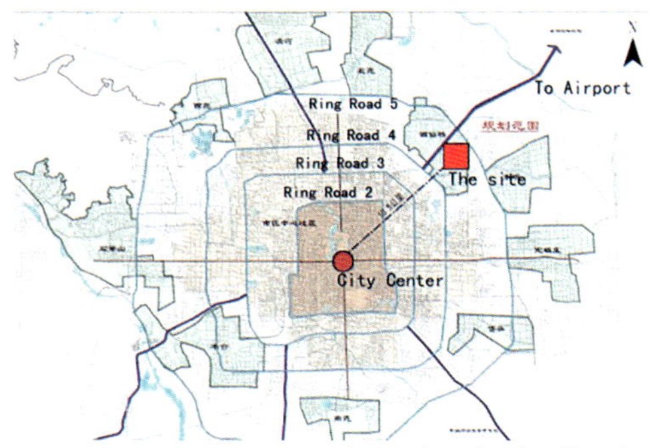

Location in the Region

Location in the City

Location in the District

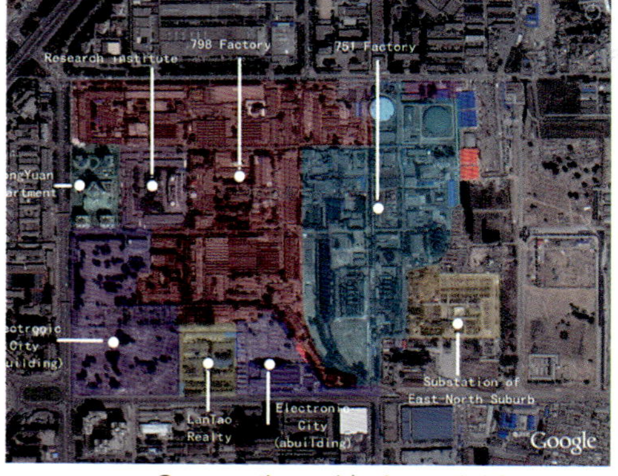

Connection with the neighbouhood

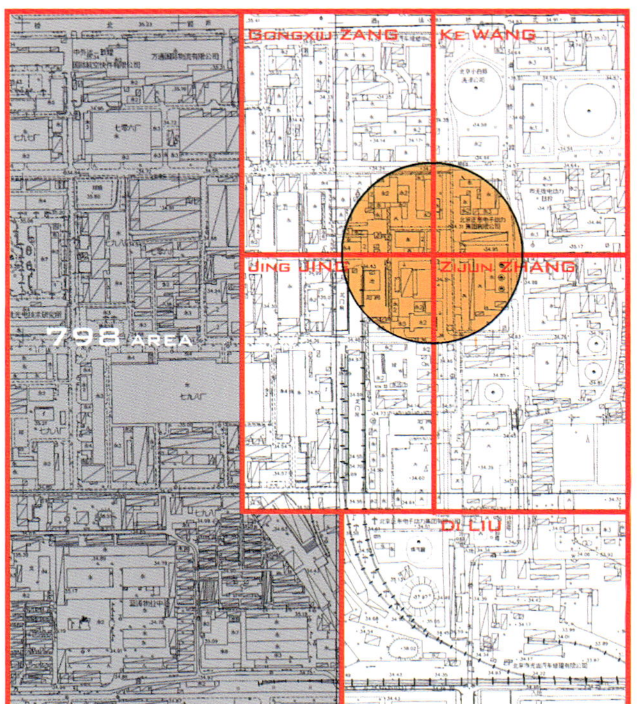

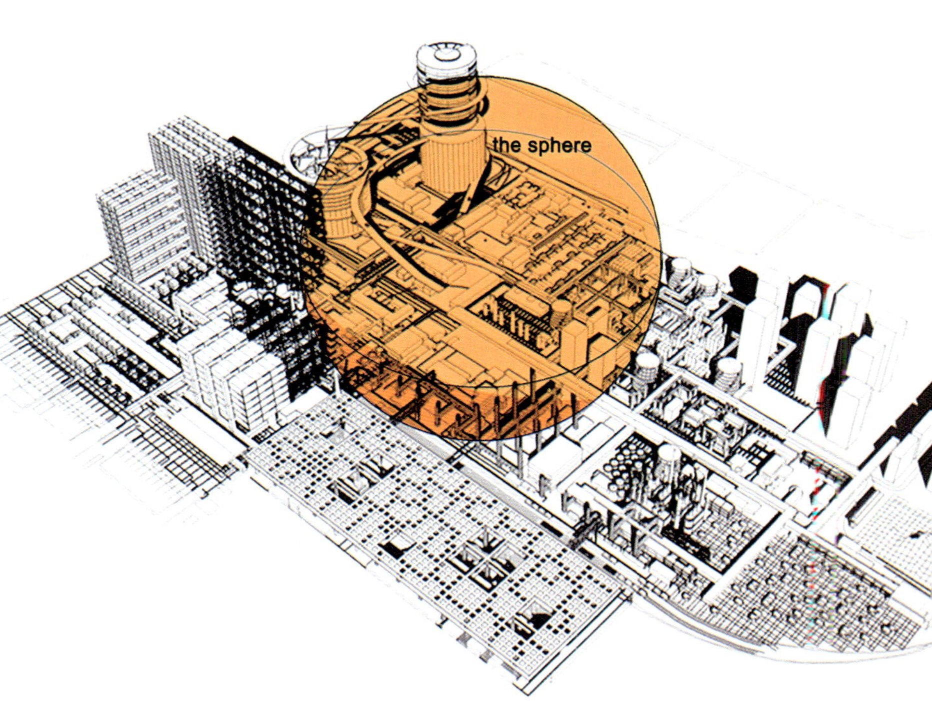

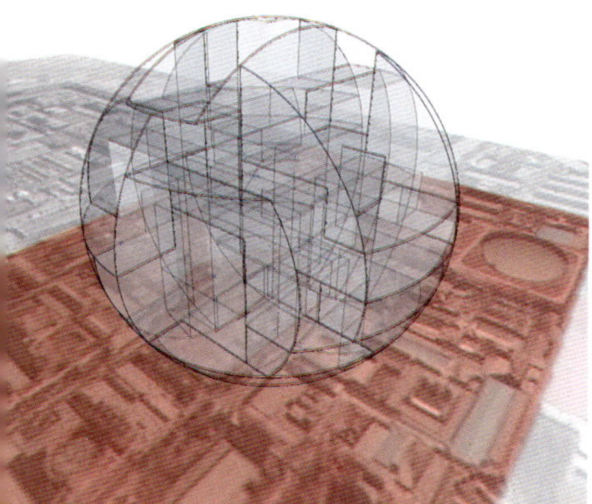

Zone 0 - area 751
零区—751地块

tutors: prof. Dong Wei, Zhang Qian, Du Rong
students: Zang Gongxiu, WangKe, Jing Jing, Zhang Zijun, Liu Di

在此次与荷兰代尔夫特理工大学的联合教学中，我院有5名五年级的同学参加了课程设计，3名教师进行辅导。与代尔夫特的同学们不同，我们不是在一个三维的球体中切分体块，而是在一个二维的平面上切分用地、展开设计。为了鼓励交互式合作设计的展开，我们随机地在用地中间进行了切分，而不是常规地沿路划分每个同学的地块。此外，我们还面临着相对实际的问题，一是老工业基地的保护；二是由首规委提出的，能否实现一个较高的密度。

751地区过去是一个能源动力工厂区，未来，将随着毗邻的798等地区的改变，更新为以文化产业为主，多种城市服务功能并存的混合街区。在11周的课程中，每一个同学都根据基地和使用者特质，设计出了灵活多变的方案，可以根据使用者的需求和使用者本身的不断变化而改变建筑体量和房间分隔。控制这些变化的都是一些非常简单的规则。与此同时，同学们也展开研究，针对保护的不同对象——建筑和大型设备分别设计出了富有创意、别具一格的方案。通过持续的网络交流和3次同步的汇报，同学们和代尔夫特理工大学的同学们保持了密切的联系；通过组内的数据交换和沟通，5名同学在保持各自方案特色的同时，实现了总体功能指标的平衡。

The rest of the 751 area, around the virtual sphere, has been the subject of a parallel design studio held at the School of Architecture at the Southeast University in Nanjing. Different from TUD's three dimensional sphere, the two dimensional area around it was divided into 5 parts. The dividing boundaries were always within the plot instead of along the roads, which encouraged the collaborative design. Many more practical tasks had to be solved, such as the conservation of old factory buildings and a higher density required by the local government.

Formerly a Bauhaus-style energy sources plant, 751 confronts the regeneration to be a multifunctional culture district. During the 11 weeks of design work, each SEU student developed a flexible design which would suit the diverse life of artists and other users. Controlled by several simple rules, the blocks or the rooms in every design can be changed easily according to the users demand. At the same time, they created distinct and exceptional schemes to conservation the old buildings and facilities. Data change, presentations and contact with TUD students helped them in collaborating with each other and balanced the functions at the whole site.

Carrying Belt:
A belt carrying coal from the train station to factories.

antry Crane:
g machine has e shape like the or. Used for rry heavy things.

Air-cone:
A group of machines used for making natural gas.

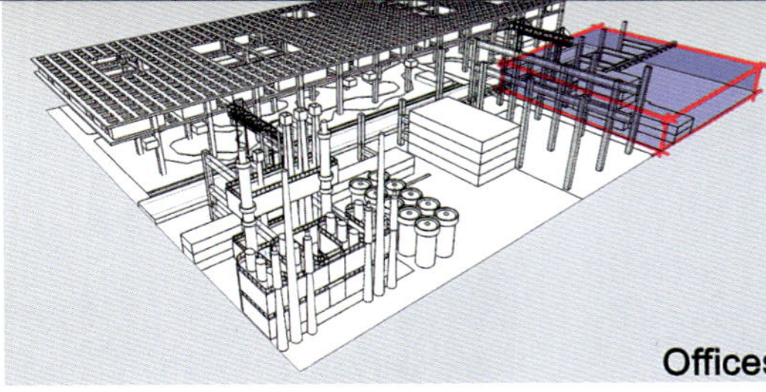

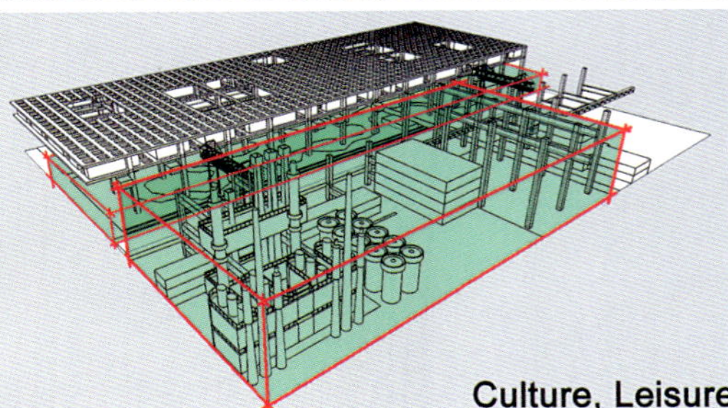

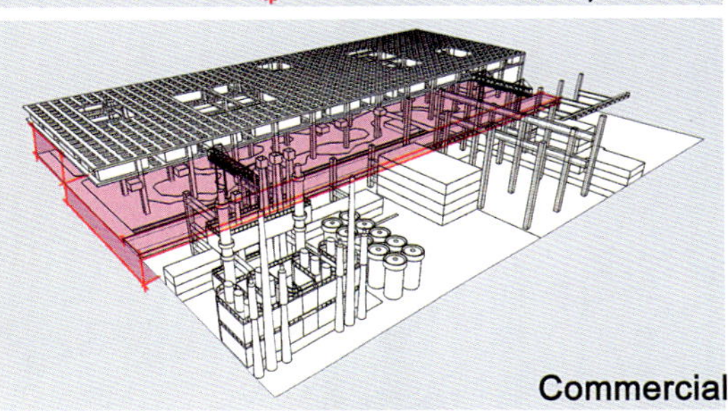

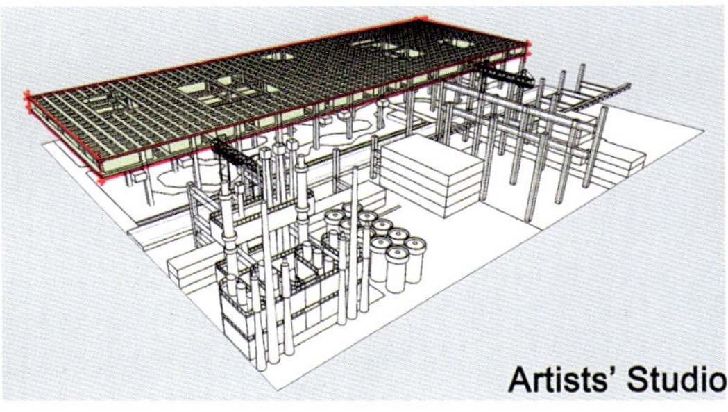

In my opinion the artists should be the GOD without any questions. It's the artists who first rediscover this area. It's the artists who bring energies and fresh-air to this area. It's the artists who make this area known to all. It's the artists who make the society rejudge the area. It's also the artists who renew the spirit of this area. And I am sure that the artists will keep on doing their contribute the area.

经过分析，我选择了艺术家这一群体，因为他们使这块地重新被大众关注，他们使社会重新审视这块被废弃的工业遗迹，他们给这块地带来了生机和活力，他们是这块地的精神领袖，他们是这块地的灵魂。所以，在设计中，我也考虑到他们在精神上的地位，将他们的工作室放在空中，让他们能审视整个地块，产生创作的激情。而地面将留给居民和旅游者，原有的工业遗迹（厂房、机械等）都被保留下来。

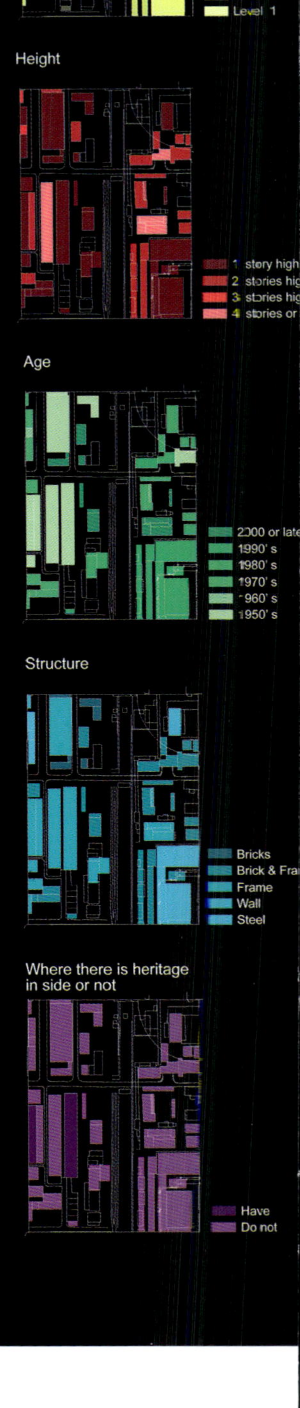

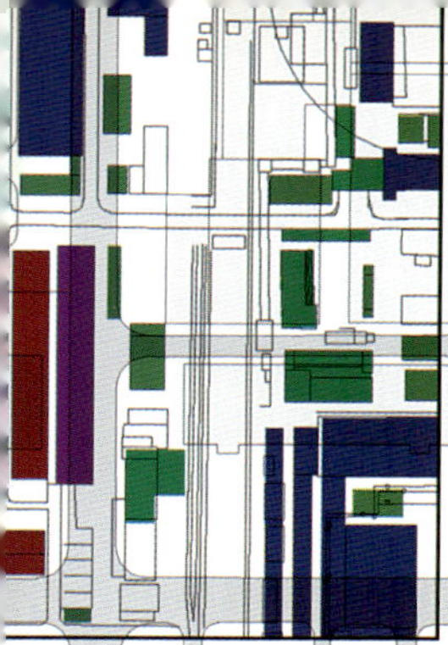

Removing
Renewing
Undecided
Keeping
Protecting

;od occupy the sky, while
oulaces acting on the
ound."
veral factors have been
en into consideration:
ality, height, age, struc-
e and whether there is
itage inside or not.
erwards, we could finally
: a general estimation of
se buildings. Five tech-
ues are being adopted:
tecting, keeping, renew-
, removing and unde-
ed, by which we could
ntinue the context of the
ustry site. Several build-
s have been combined
a huge one to create the
ibition space, and some
en-spaces also have
en created.

"上帝在空中俯瞰大地，众生在地面演绎生活。"

通过对地面建筑的质量、高度、建造年代、结构方式以及是否有文化古迹等角度分析，得出一个建筑拆建分布意向图，将地面建筑分为保护、保留、可拆可留、改造以及拆除五大类，以此为依据来保留基地中的建筑，最终保留工业遗迹的文脉。将保留下来的建筑再进行整合，将一些小建筑群改造成大空间。最终地面形成文化休闲以及商业为主，附加停车以及广场空间。

PLOT A

- A1 特色纪念品专卖店
- A2 特色餐饮店
- A3 艺术品展示馆
- A4 旅游文化广场
- A5 知名艺术家画廊
- A6 休闲活动广场
- A7 人工湖
- A8 半地下酒吧街
- A9 龙门吊改造运输设施

PLOT B

- B1 物业管理楼前广场
- B2 物业管理楼
- B3 空中步行传送带
- B4 大球体接地开敞空间
- B5 地下酒吧前集散广场

PLOT C

- C1 商业广场
- C2 艺术家作品专卖店
- C3 商业步行街
- C4 艺术品展示馆
- C5 特色餐饮综合体
- C6 露天咖啡座
- C7 地下停车场
- C8 人文教育广场
- C9 地下储藏空间入口
- C10 地下商业酒吧广场采光顶
- C11 休闲活动广场
- C12 半地下酒吧街
- C13 地面停车场

PLOT D

- D1 地面停车场
- D2 地块间开敞空间
- D3 空中步行通道
- D4 垂直交通体
- D5 工业遗迹露天展示
- D6 工业遗迹展示馆
- D7 储油罐改造休闲体
- D8 地下酒吧前集散广场
- D9 重油裂解炉改造综合体

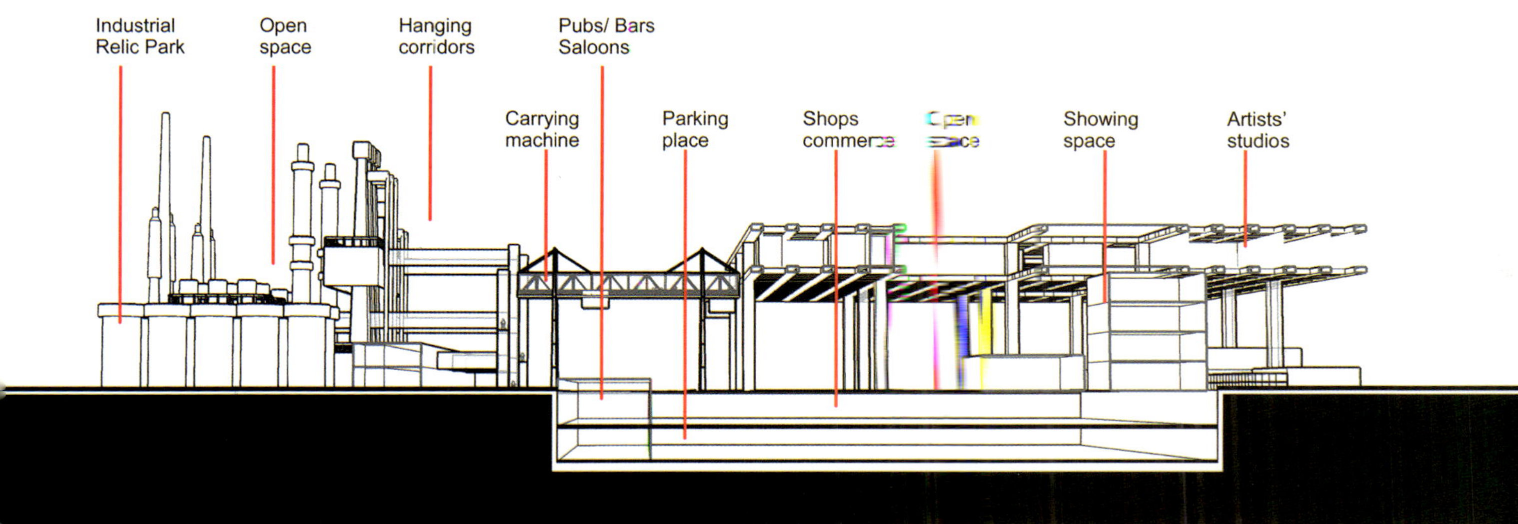

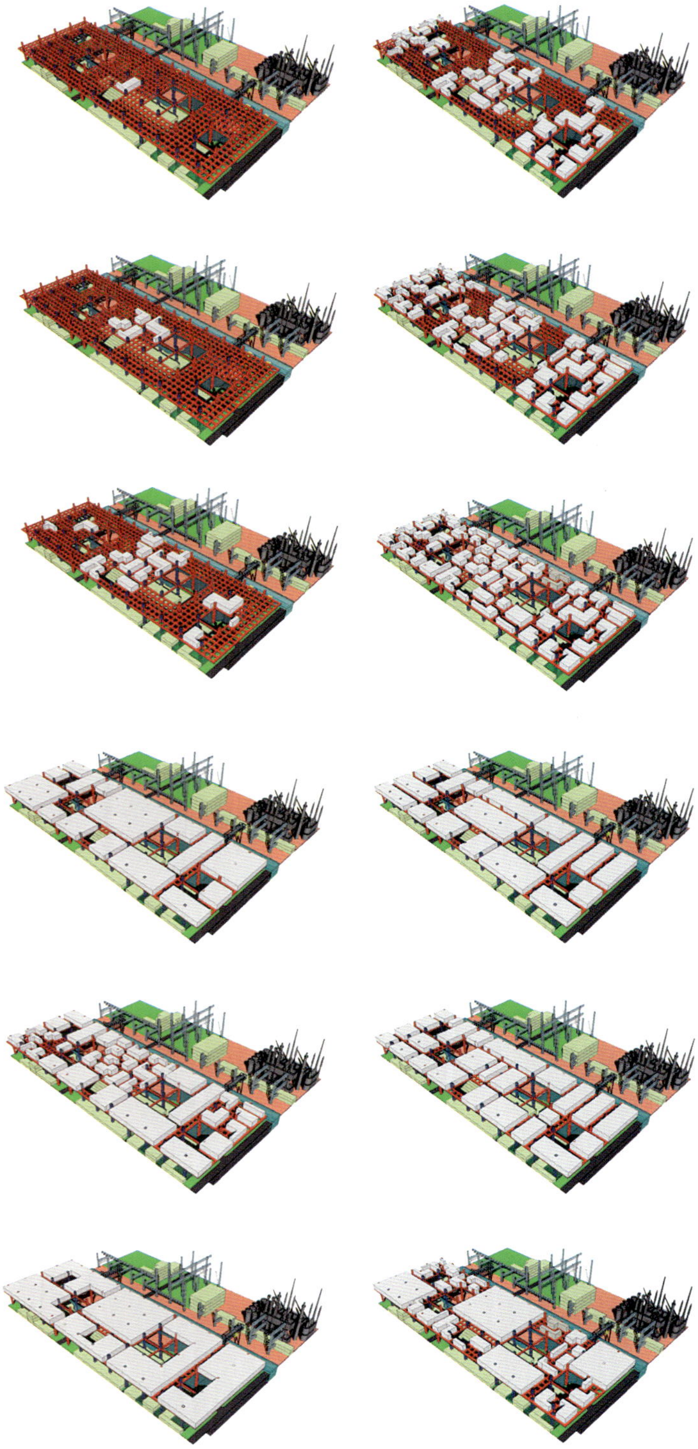

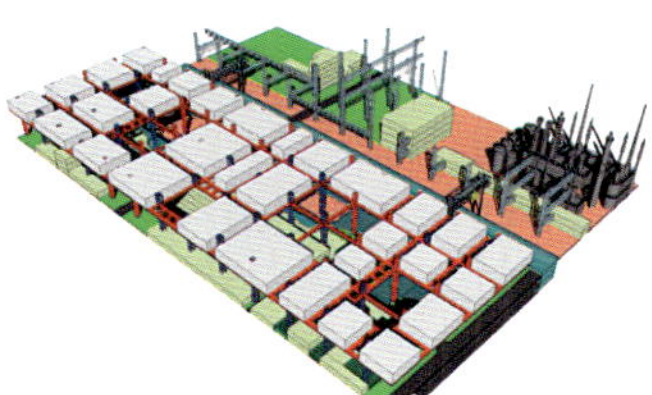

Hanging Passway: also functions as the beam. 2m width, space between: 4m.
Sustainable structure: space between: 25m. Functions as the elevator when there is a building on the ground.
Artists' Atelier: storey height (can be divided into two storeys): 6m; Size could be adjusted accordingly. There is no atelier above the existing buildings in order to ensure the sunlight; The newly built atelier should be one grid away from the existing ones.
Auxiliary block: Position and quantity can be adjusted according to the number of artists.
To satisfy different people's dissimilar request, the studios' shape is not always the same, as well as the size. Sometimes it can be small, and sometimes it can be big, it can be long and narrow or capacious. And as a result the form of organization of them is different.
Sometimes if a big company comes, maybe a huge space would be needed, but the already existed studios already take their position, and divided the whole area into several pieces. It may have to talk to them to make an agreement to reoccupancy the area.

空中网架：宽2m，间距4m，同时起到结构作用，支撑空中艺术家工作室。
支撑结构：2mx2m见方，间距25m，支撑起整个空中往架。同时，有一部会支撑结构，当其落地到建筑上的时候，又起垂直交通的作用。
艺术家工作室：层高6m（可以形成夹层），需艺术家自己搭建，形状大小也由艺术家自己设计。但是在搭建过程中必须满足一定规则：地面建筑的上空不能建造；后进入的艺术家需要考虑先进驻的艺术家，也就是说，两个相邻工作室之间的间距要有一个网格，这样可以满足采光以及一定的私密性的要求。
辅助体块：悬挂在空中网架的边缘，提供一些辅助功能，如杂货，可以根据艺术家数量以及密集程度调整其位置。
这样的设计可以适应不同需求的人群：当入驻的都是小型的艺术家群体的时候，可以是一种景象，当有大型的社会团体进驻的时候又是另一种景象。这样的变化多端，可以使整个基地充满生机和活力，可以避免一尘不变的尴尬。
人们总是喜欢在阳光下活动，因此，空中网架不能是满铺的网格，我在网格上留有一些大的空洞，这些洞在地面形成的巨大光斑对应的就是人们主要活动区域—地面广场和湖中的主要岛屿。
空中交通：原有龙门吊改为自动传送器，利用其原来的水平位移和垂直起重的作用，利用到交通上，可以水平和垂直传送人。原有运煤的传送带改造为步行传送器，利用其负重传送的原理，应用到交通上，可水平传送人。它们再和垂直交通体一起，构成整个空中交通，使整个基地东西两片区，以及空中地面都成为整体。
地面交通：将原有路网整合，连通798和751，和其他分地块的地面道路对接，再在主要出入口附近设置停车场，满足一部分的地面停车要求，最后通过步行交通联结广场。
地下交通：地下二层是停车场

We prepared a frame and several boards, which can be used as the walls, floors or ceilings, and by using these materials, one can make a studio and decide its shape, area etc. by oneself...

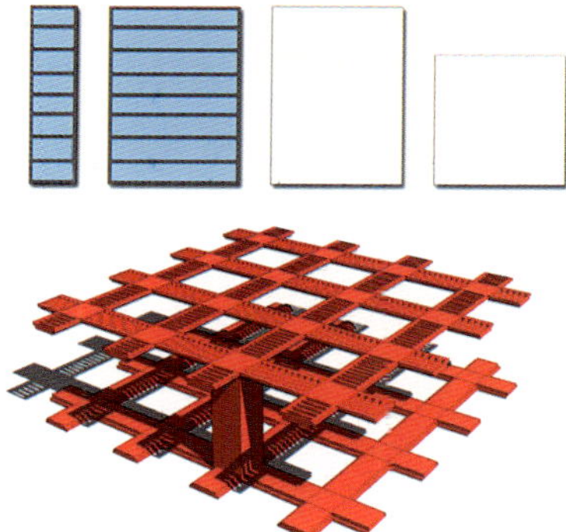

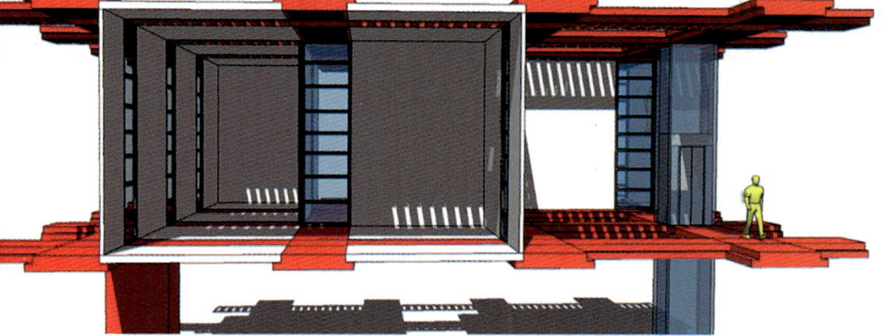

If you want big space you can make your studio like this.

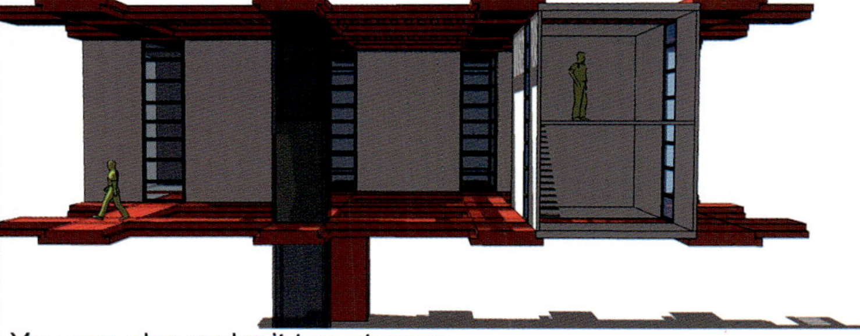

You can also make it two storeys.

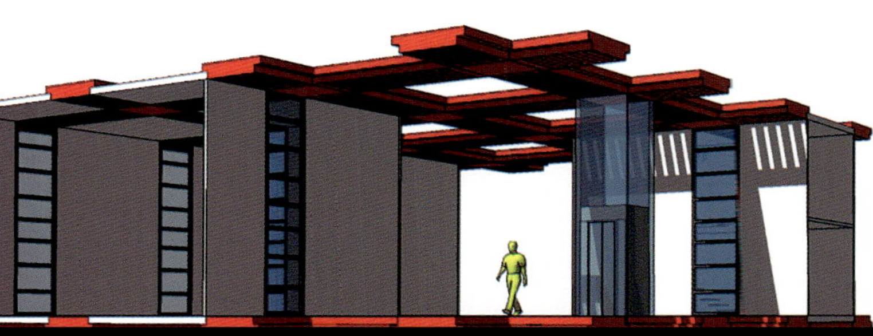

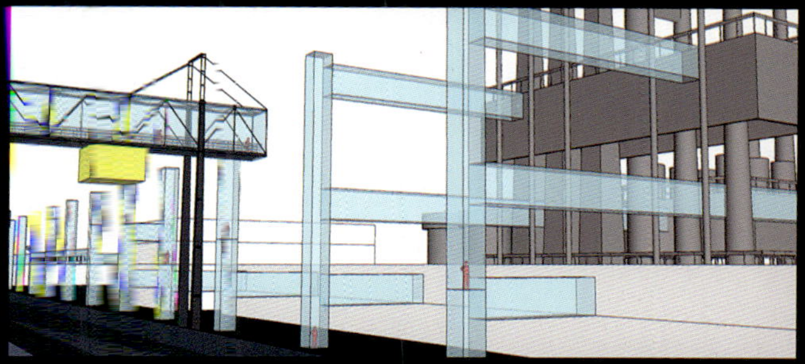

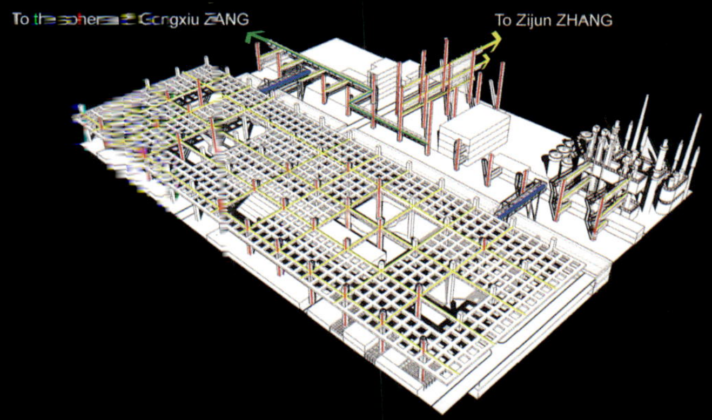

To Zijun ZHANG

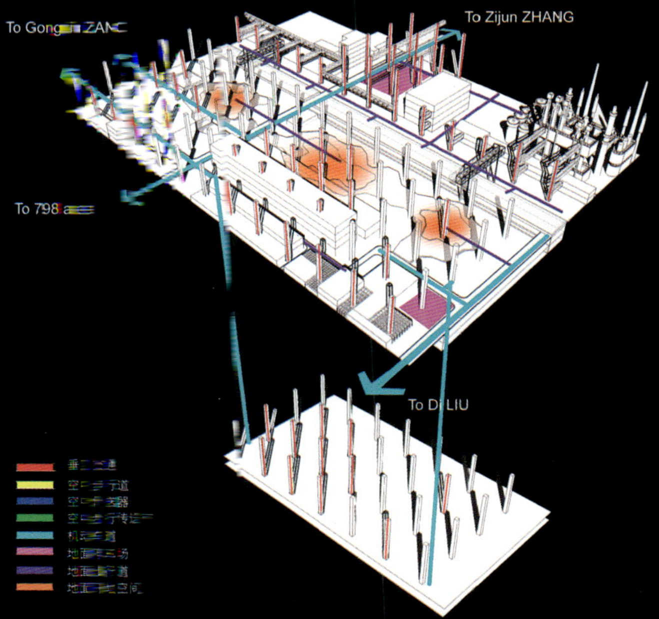

To Zijun ZHANG

Location
基地

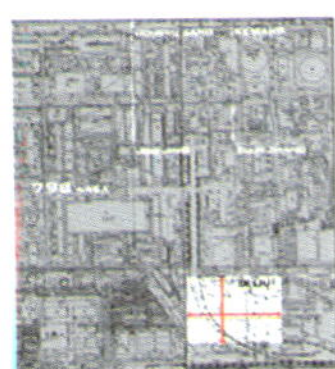

元素1: 重油裂解炉[Element 1: the large oilcans]

个人方案

地段外沿存在两个旧工业遗留下的重油裂解炉规模巨大且原貌完好，考虑保留作为原址工业的见证

There are two large oilcans in the location.In my project they are reserved.

[现状及改造]西侧重油裂解炉规模巨大，设备庞杂。各种管道之间分割而成的空间极为丰富。改建将在原基础上加入多层排架结构，与原有结构相结合使上层空间得以利用。穿插大小不同的艺术工作室。

[Actuality and Rebuilding] The rebuiding to oilcan in the west(A) will make use of the excrescent steel gridding to creat various spaces in it as below.

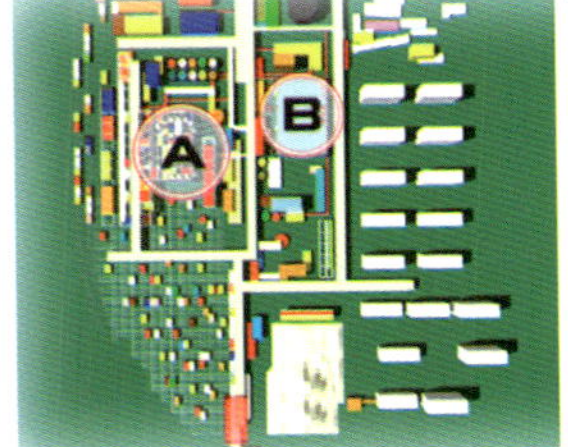

[现状及改造]东侧重油裂解炉在规模，跨度，设备数量方面相对西侧均较小。裂解炉之间的间距较大。考虑利用其内部大空间改建为展览馆，采用半透明玻璃体外皮，使得新建筑不仅具有现代的材料与构建方式，同时透明的材质又可将历史上的工业遗迹得以体现。改建后可作为工业历史展览馆。

[Actuality and Rebuilding] The oilcan in the east(B) is a small one. Here will make a cover outside of it with glasses in order to show the bequest during the industry age and it will be a industry museum in the future.

元素2: 架空管道

改造考虑利用架空管道将北侧电厂余热向周边地区供暖。具体方案为在参与供暖管道加入空中封闭步道。

Here will set a stepped space outside of the existent pipes. There will be also making use of the excrescent energy of the power house in the north of the base.

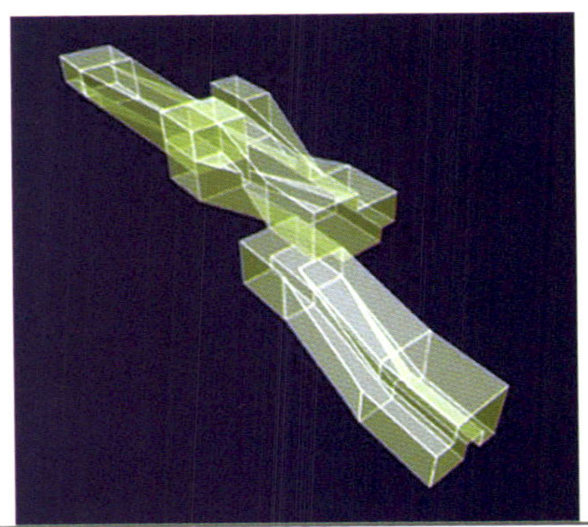

元素3：滑动单元－ 艺术家工作室

为了实现单元的可动性。引入Virtools工具，用于艺术家们对自己工作室在空间上的控制。
地块南端结合现有运输轨道，利用保留下来的厂房框架结构形成网状的水平与垂直滑动模式。新加入的滑动单元尺寸为4m×6m，两个或多个单元之间可以合并组合。但需要遵守以下原则：

With making use of the rail in the location, the unit for the artsts which is designed as 4m×6m would be able to slide on if only obey the principle bellow:

滑动组合原则：
原则1
将每两个单元之间的碰撞属性改为3m内排斥，
使任意两个单元之间的距离将不得小于3m

原则2
当需要大空间时，点中任意两个或多个需合并单元，
右击退出原则1，同时进行合并，之后返回原则1。

合并好的较大单元又将以整体参与上述原则，
与其他单元保持3m的距离。

注：关于Virtools：该软件的基本操作原理即对计算机输入一定的变化运动原则后，计算机将会通过程序计算来实现对物体的变化操控。

Principle 1
Keep the units in the distance of 3 meters.

Principle 2
Quit principle1,unite two or more units when there is needs.then back to principle1.

The united which had been a bigger unit would obey the principle1 too.

NOTE: The Virtools will complete the operation once we did as above.

BLOOK -2空中通道Air-road

王柯 Wang Ke

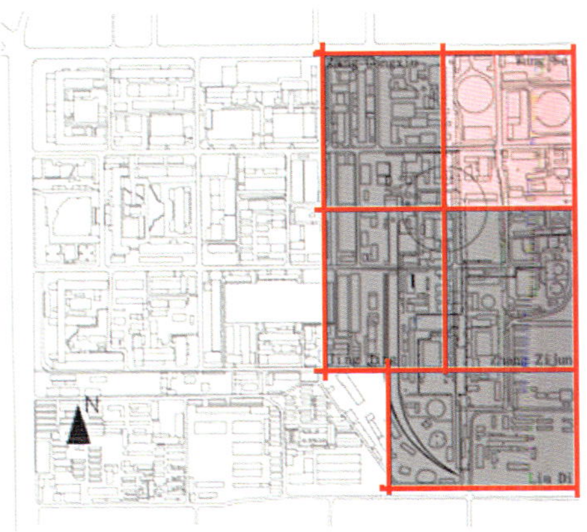

设计意图
Design intentions

该场地上最引人注目的是两个高60m的煤气储罐，而且它们体积可以根据储气的多少进行变化。因此如何利用这两个高大的储气罐将成为设计的重点。

The site is remarkable for the two high(60meters)and big gas-hol-ders which could change their volume by the gas amount. Then how to use these gas-holders is the important point in our design.

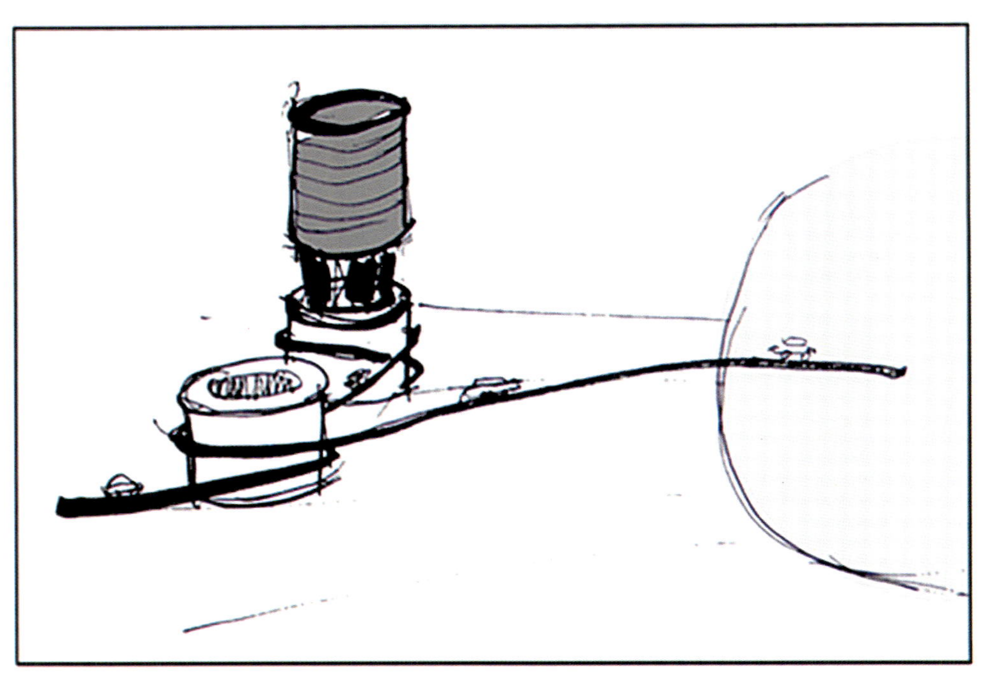

煤气罐和球体、煤气罐和加建建筑的关系
Gas-holders and Sphere additional bulding

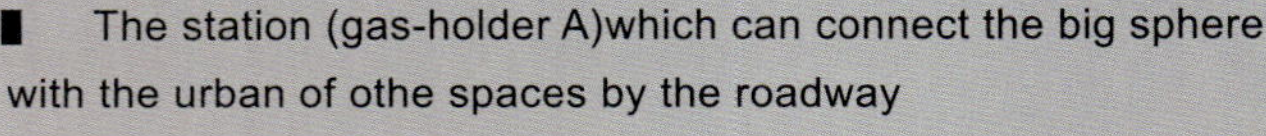

- The station (gas-holder A)which can connect the big sphere with the urban of othe spaces by the roadway
- The additional building above the gas-holder B which is conn ected with the urban or other spaces by the roadway

车辆入口
Vehicular entrance

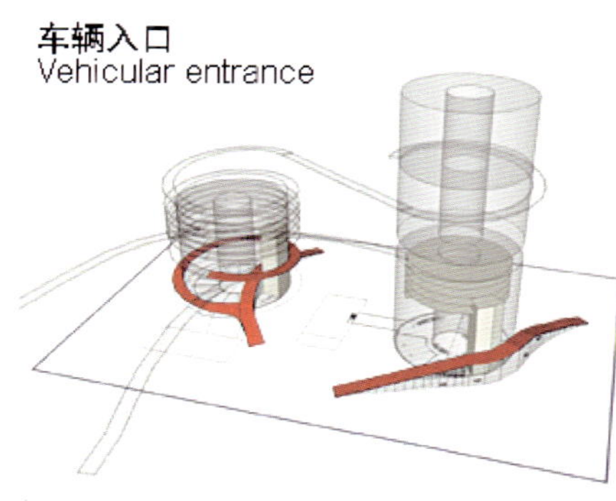

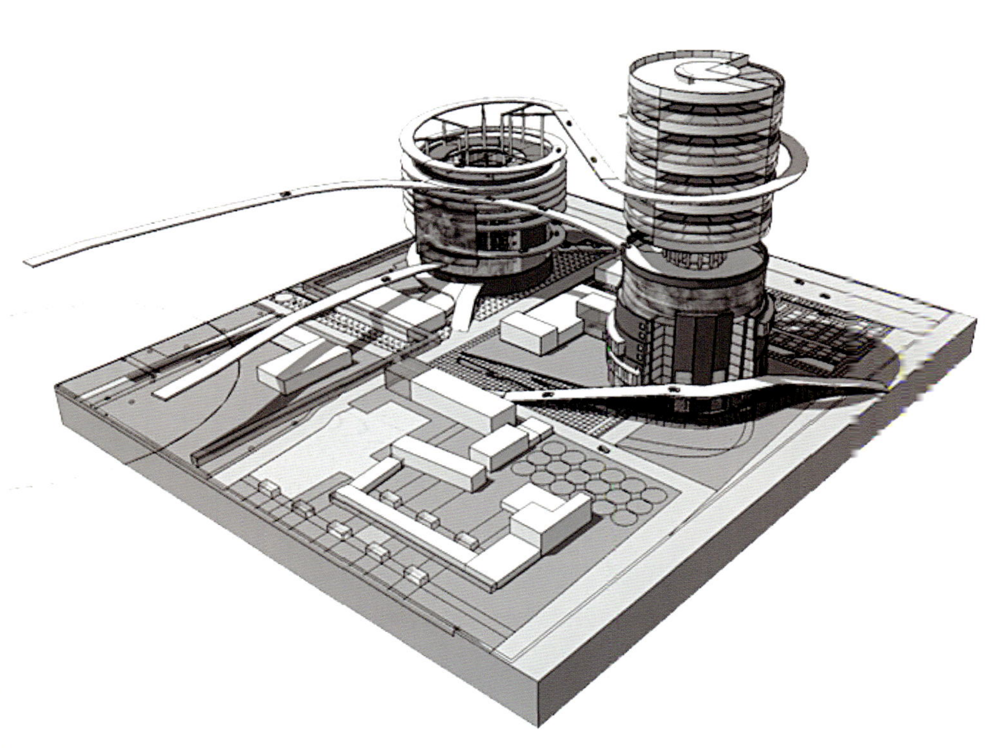

垂直交通
Vertical traffic

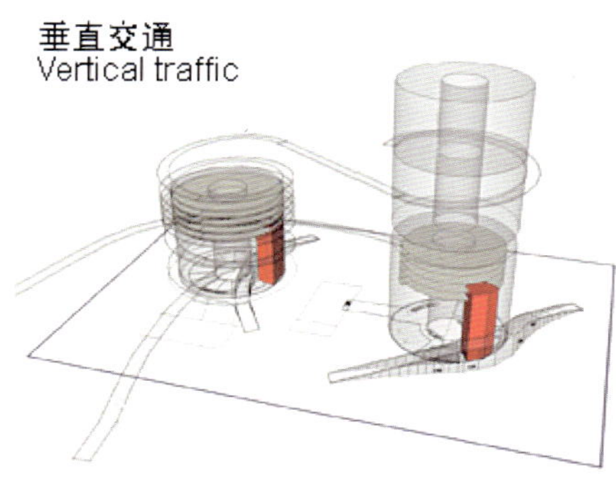

车库出入口
Garage entrance

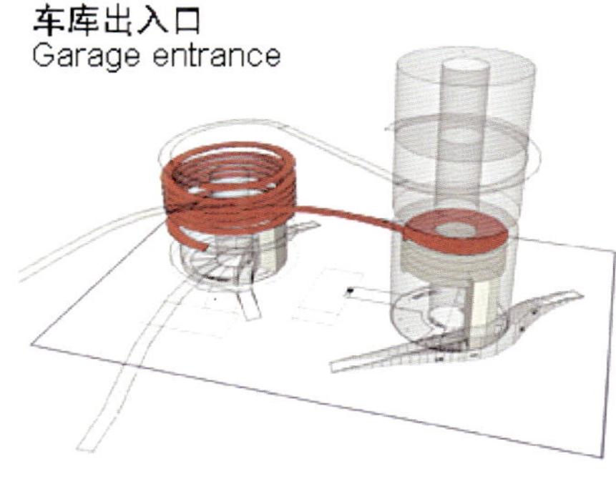

连接通道
Connections

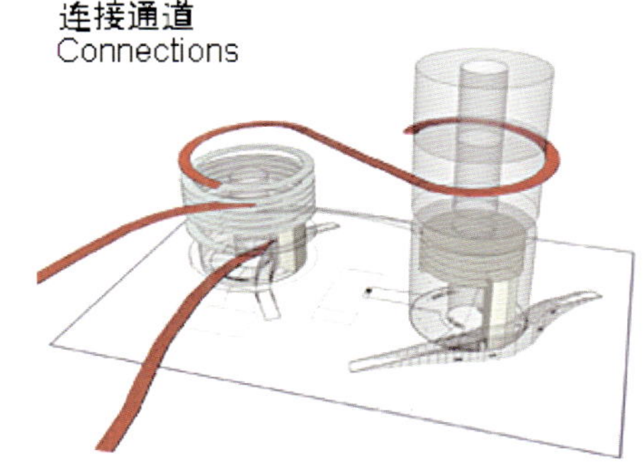

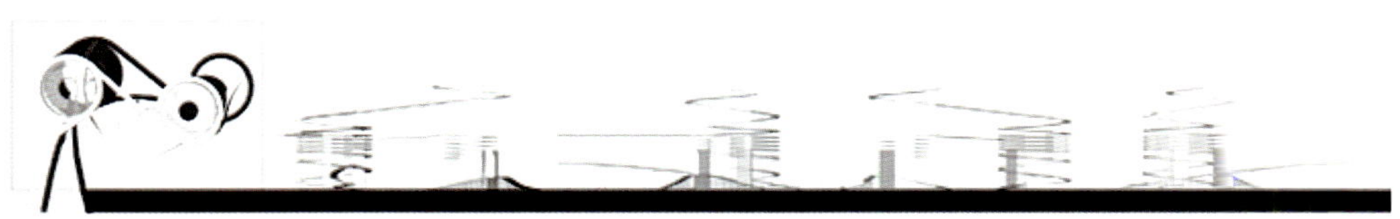

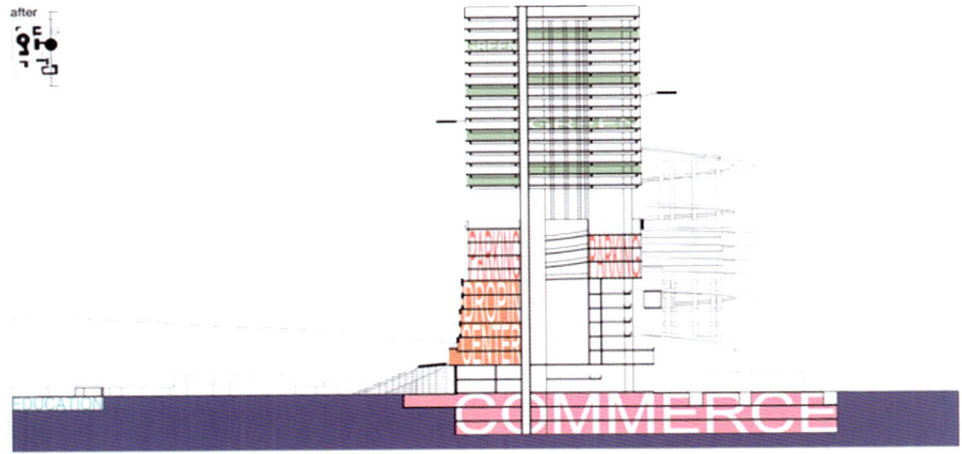

东边的储气罐有两面面向城市，于是将它改造为具有混合使用功能的公共建筑（顶层及地下两层作为停车库），并在其上加建了一个商住一体的建筑。西边的储气罐将作为大的球体与城市的机动车转换站，机动车可以通过该建筑到达大球体或是进入东边的储气罐和加建部分。同时，该转换站可以利用其可变的体积改造成为根据不同需要可以随时调节的展区.

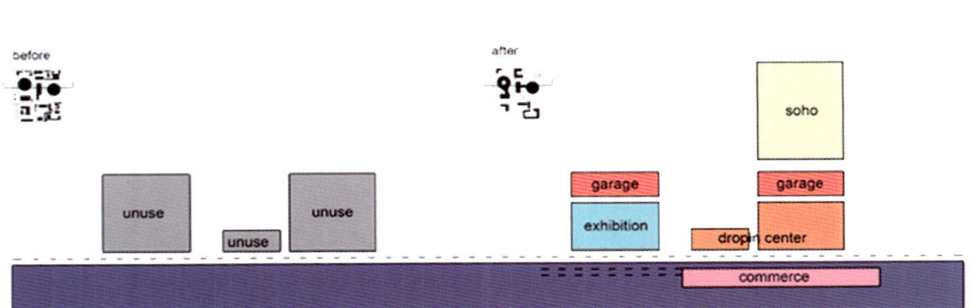

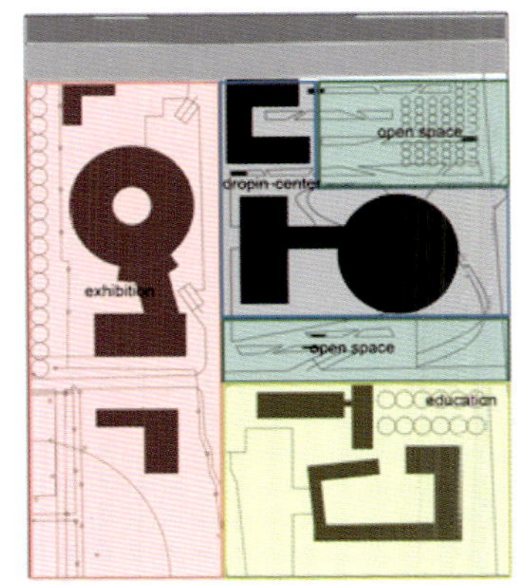

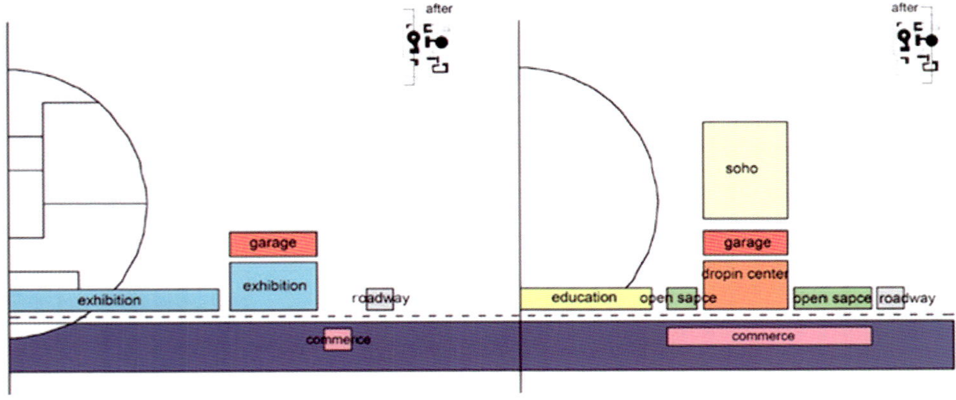

The east one has two sides facing the city,so it can be designed to be the mixed -function al public building (the top level and below is used to be the garage, and add a new building which is used for living and commerce. The west one will be the station which can connect with the big sphere or the east buildings. Meanwhile, this station can be used to be the exhibition hall.The hall volume can be changed any time because of the changeable gas-holders.

BLOCK -3 管道系统The Pipelines

张子君 Zhang Zijun

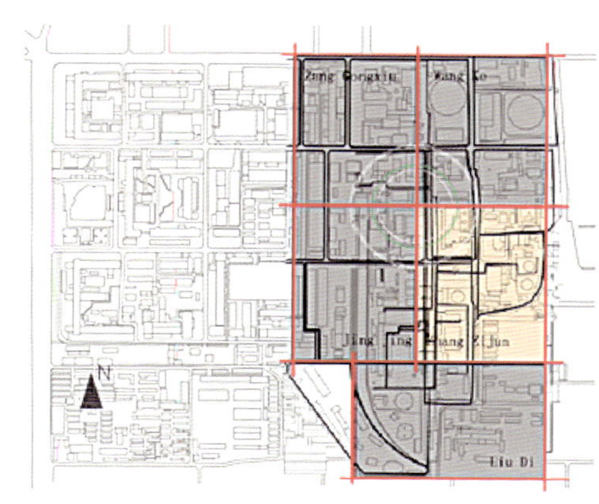

设计意图：
Design intentions

我的基地位于751工厂的东部，该场地上有很多管道和以前在工厂中很重要的大罐子。

My site is on the middle east of the 751 factory.There are many pipelines on the site.And the big boiler is the most important things in the past of the factory.

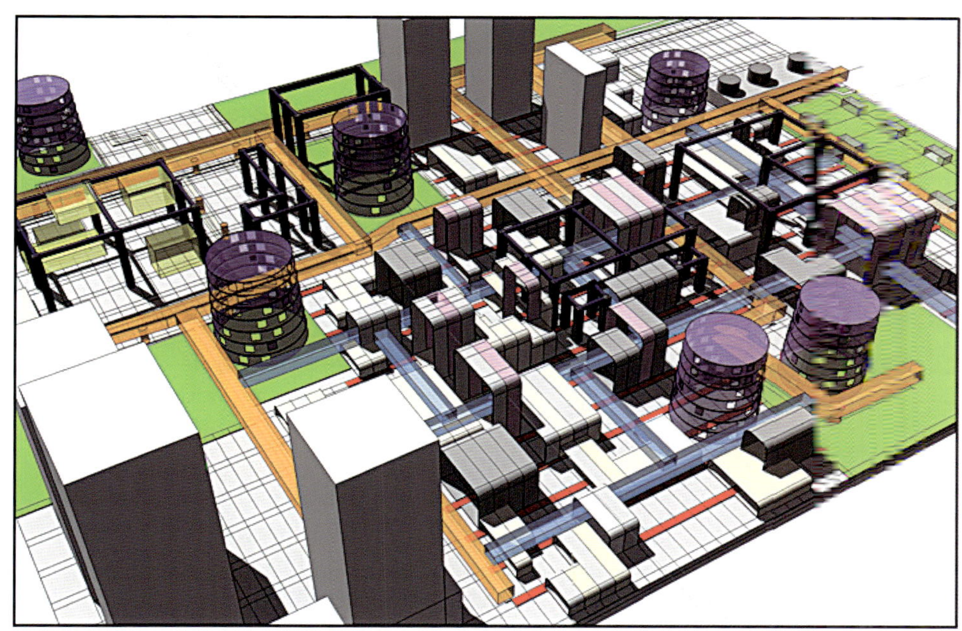

整体意向
Integrate intention

管道式基地内最主要的组成元素. 在设计中, 将这些管道重新利用, 成为空中清晰的道路系统, 直接通到各个独立的艺术家工作室.

The most important elements in my site.So we rebuild the pipelines and use them as a walking channels. It become the clear road in the air. If people want to arrived to anywhere, they should use the pipelines.

主管道：

和大罐直接相连，最主要的交通路经。
管道大小：4m×7m

The main pipelines:

The main pipelines are connected with the boilers directly,
The size of the main pipelines: 4m×7m

次管道：

二级通路，连接主管道和整个艺术家工作室的次级道路系统。
管道大小：4m×4m

The secondary pipelines:

The secondary pipelines are connected with the main pipelines,
The size of the main pipelines: 4m×4m

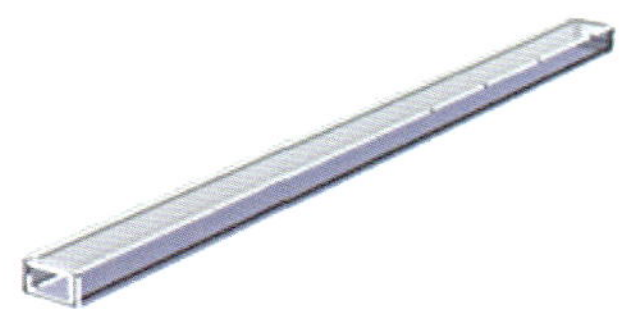

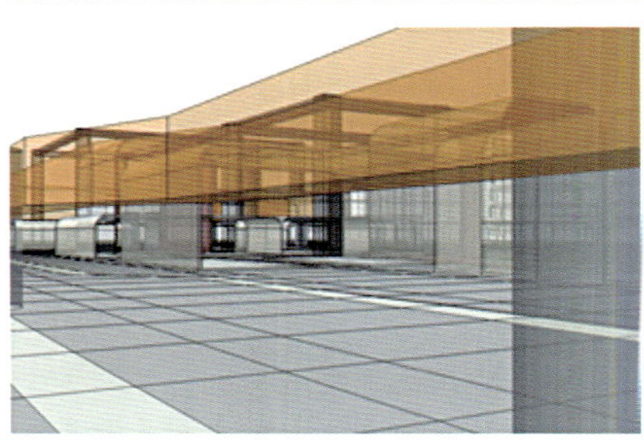

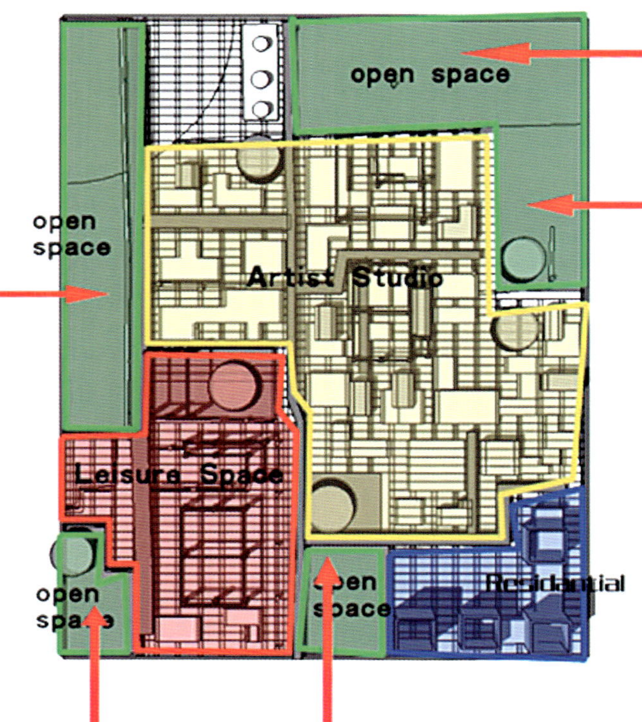

功能分布+出入口

基地由四个功能区组成：
1. 工作室
2. 公共休闲空间
3. 开放绿地空间
4. 居住空间

The site contains 4 parts:

1.Artist Studio (50%)
2.Open Space (20%)
3.Leisure Space (15%)
4.Residential Room (15%)

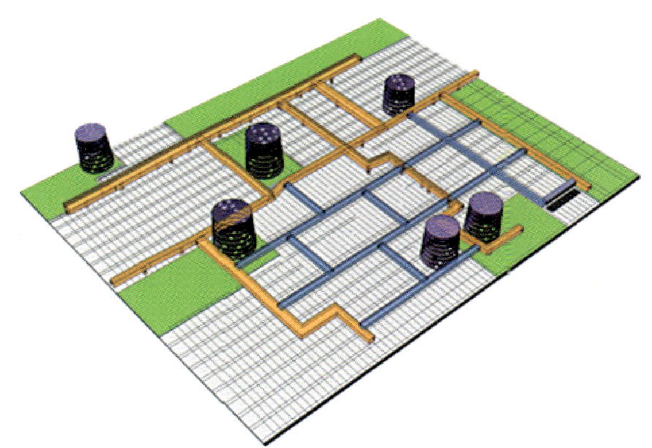

交通中枢（大罐）

基地中的大型储气罐是另一个重要元素。设计中，充分利用这些大罐的大跨度框架结构，将其改造成交通综合枢纽。每个大罐包括了：停车/交通换乘/商业/服务等各种综合功能。同时，每个大罐还都同一个绿化空间相连，具有良好的环境和开阔视野。

THE BOILERS

The symbolic elements in the site.We use them as the the "centrum system" in the design. They contains PARKING / SHOPPING / TRAFFIC TRASFER / SERVING functions. And each cylinder is near the open space on the site. The sphere which designed by TU-delft is the biggest "centrum system".

艺术家工作室设计生长规则：

1．每个工作室都必须有空间通过此管道，与之相连。
2．两个工作室东西向最少间隔4m。
3．每个工作室南北向60%的部分需要满足日照间距1：1，剩余40%部分可以随意连接，形成围合院落。

1. Each studio must cover the secondary pipeline.
2. The least distance between unlighting face between two studio in south-north way is 4 meters.
3. The 60% parts of each studio must satisfy the sunny distance 1:1 (the high of the room: the distance between the north studio and the south studio.)

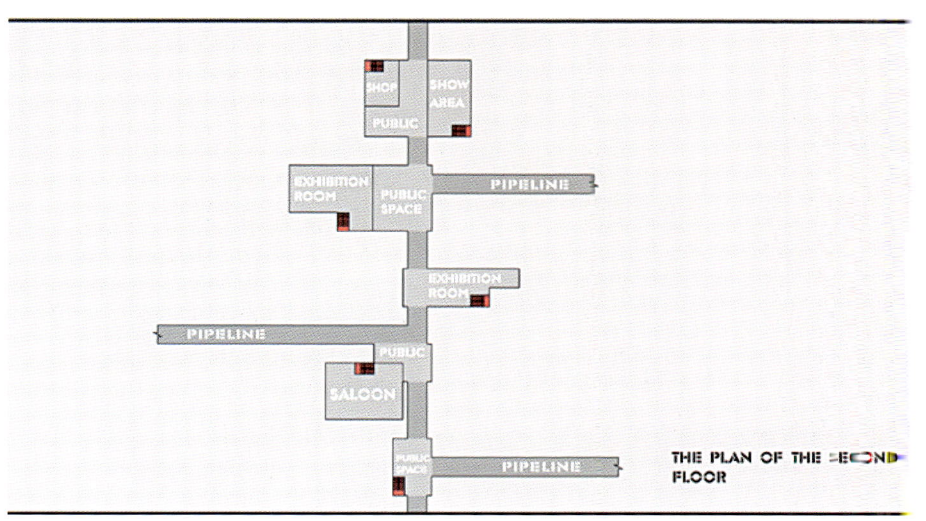

二层平面

根据管道的高度，将二层空间设置为较为公共的艺术间展示、交流活动空间。

Second plan:

Acording to the high of pipepli nes ,the second plan space is used to be the public art exhibitions and comunication sapce.

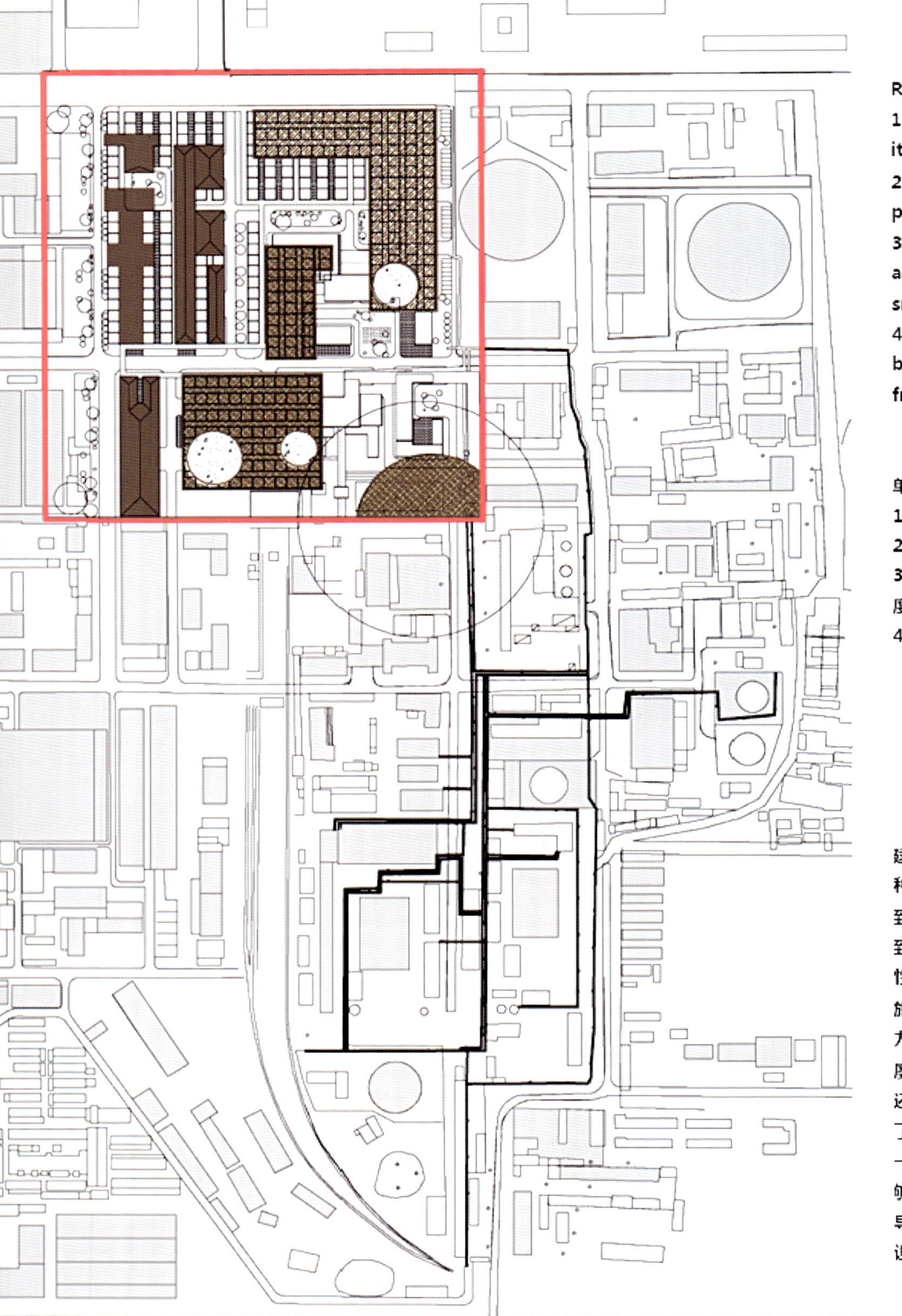

Rules

1. The unis will not gather together until its temperature is lower than 12℃.

2. The units will not depart until its tem-pera -ture is higher than 30℃.

3. The walls would be stretched when an asrtist or a group of them want a small/middle scaled exhibition ot sale.

4. The units' distance from each other wil be 7.5m from south to north and 4 meter from east to west.

单元运动规则

1. 单元只有在气温低于12摄氏度时会自动聚集
2. 单元只有在气温高于30摄氏度时会自动分开
3. 单元侧壁在单个或群体艺术家需要一个中小尺度的展览或销售空间时可以延展或收缩.
4. 单元最小间距为南北间7.5米,东西间4米.

建筑设计师处理空间关系的艺术.但是,当为这种关系寻求一种唯一的解决方式,尤其是应用到建筑设计和城市和城市规划时,设计者们遇到很大的限制.我们无法在设计中提供一种弹性,让它能够适应潜在的不稳定因素.在方案实施前后业主提供的要求,现实世界中的种种压力也使得建筑师和规划师们在实施前不得不屡次修改自己的方案.即使方案最终落成,它们还可能会在使用寿命的中期甚至早期就失去了应有的功能.因此对于实际项目而言,比提供一种单一,一次性的解决方式更为重要的是能够及时生成针对性更强的解决方案.这个需求导致能够适应不同的要求,具有较大适应面的设计方案.

用户参与的互动设计

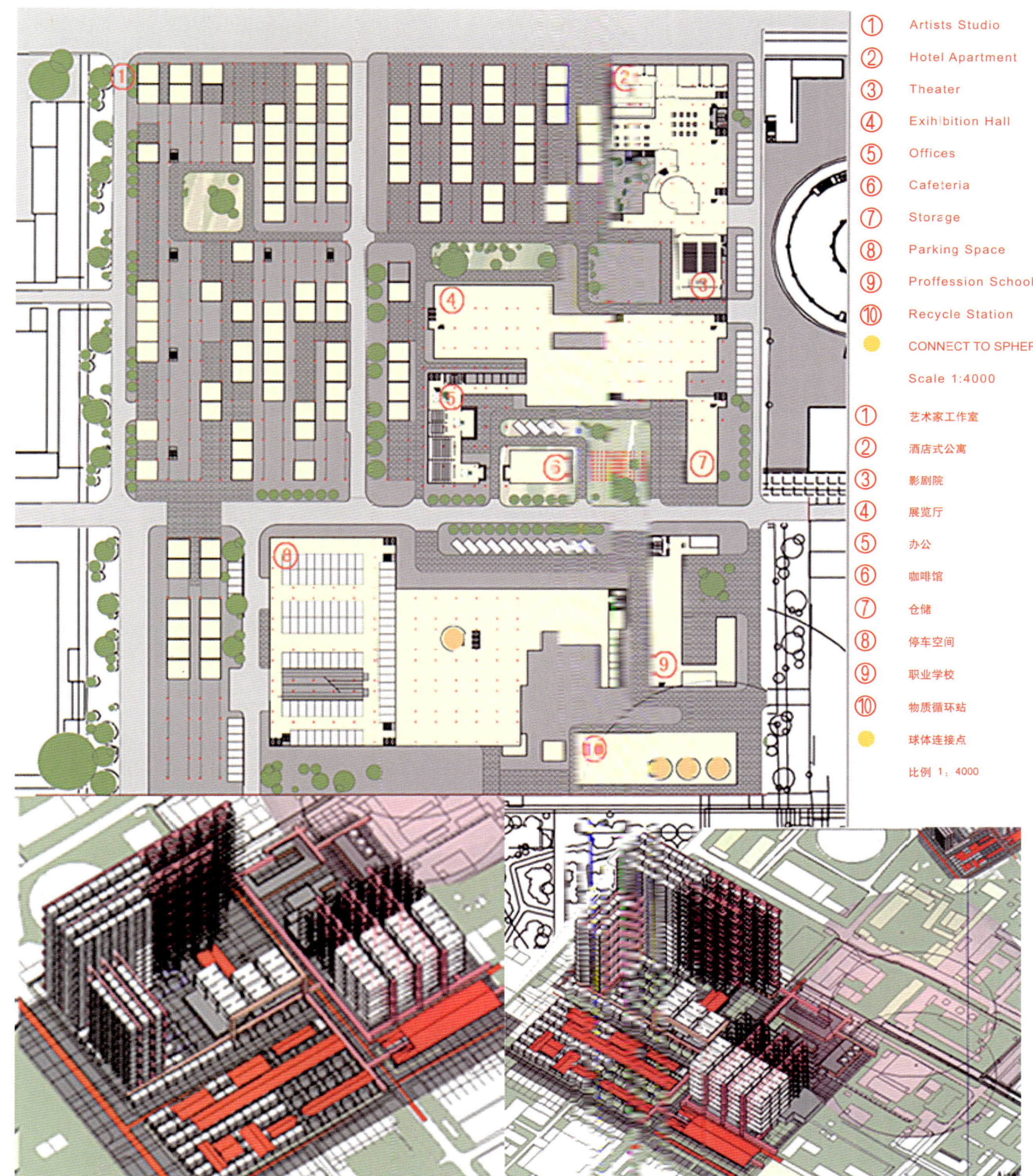
① Artists Studio
② Hotel Apartment
③ Theater
④ Exihibition Hall
⑤ Offices
⑥ Cafeteria
⑦ Storage
⑧ Parking Space
⑨ Proffession School
⑩ Recycle Station
CONNECT TO SPHERE
Scale 1:4000
① 艺术家工作室
② 酒店式公寓
③ 影剧院
④ 展览厅
⑤ 办公
⑥ 咖啡馆
⑦ 仓储
⑧ 停车空间
⑨ 职业学校
⑩ 物质循环站
球体连接点
比例 1：4000

Site Plan
总平面

North Elevation
北立面

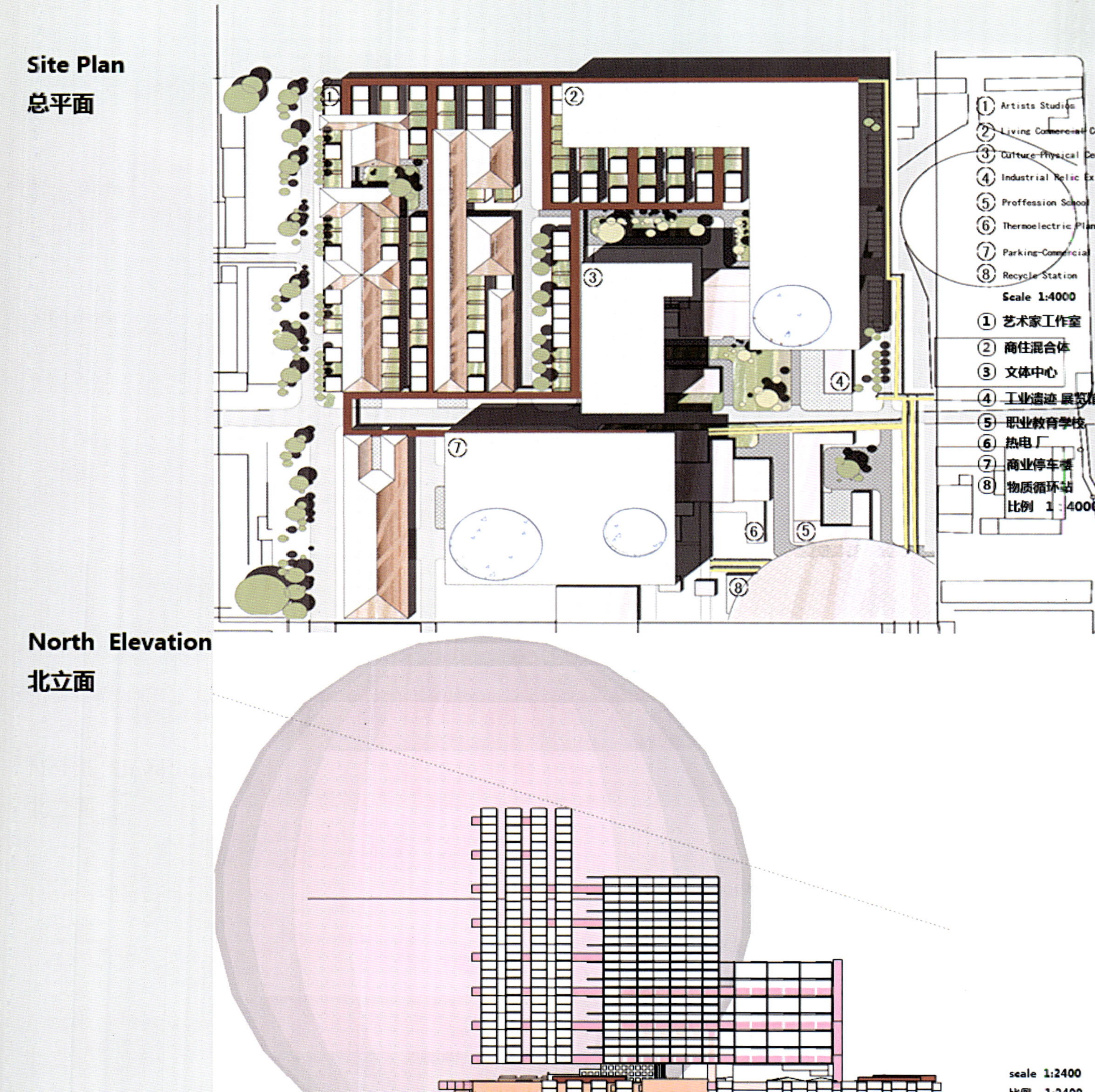

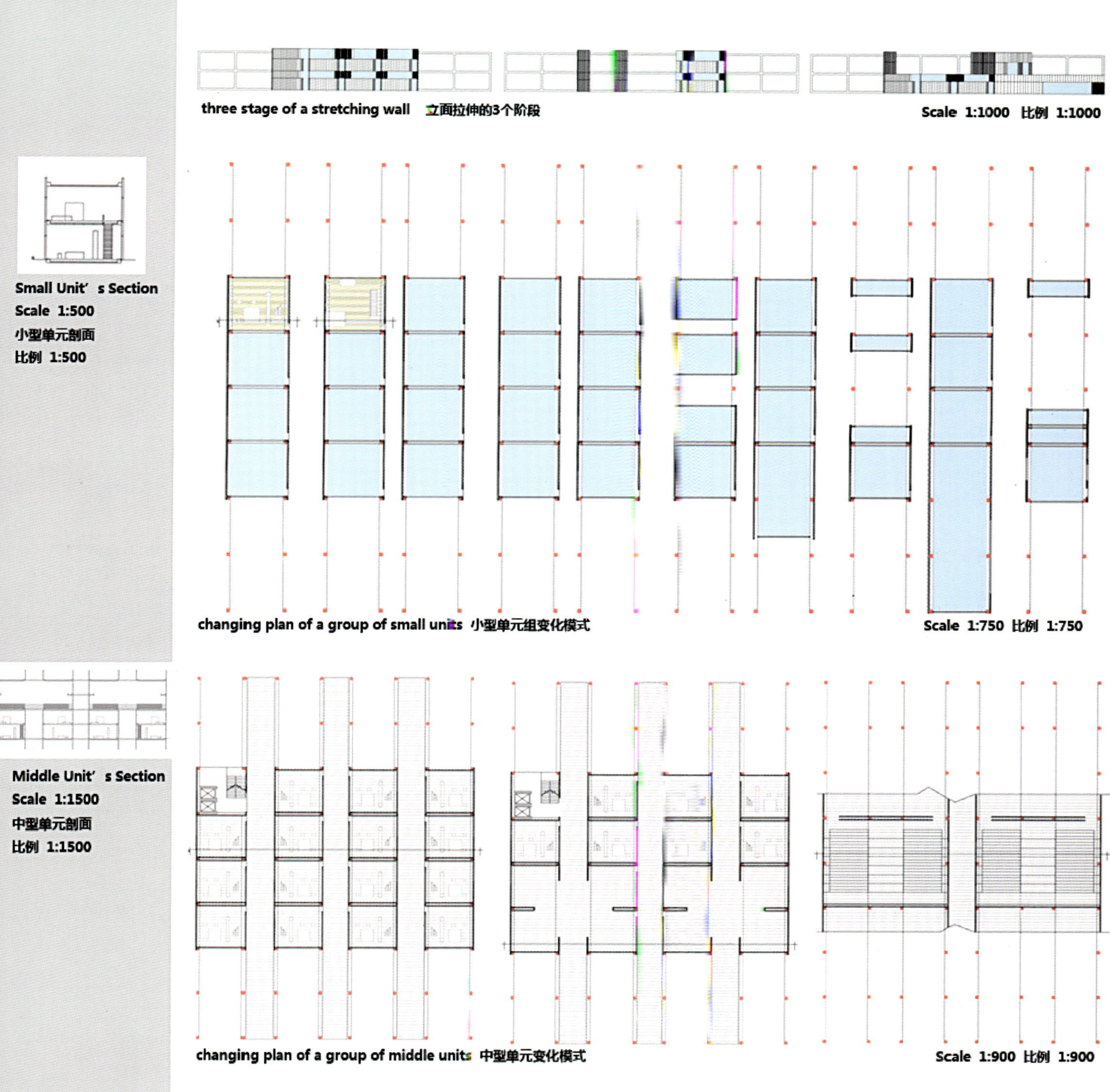
three stage of a stretching wall 立面拉伸的3个阶段
Scale 1:1000 比例 1:1000
Small Unit' s Section
Scale 1:500
小型单元剖面
比例 1:500
changing plan of a group of small units 小型单元组变化模式
Scale 1:750 比例 1:750
Middle Unit' s Section
Scale 1:1500
中型单元剖面
比例 1:1500
changing plan of a group of middle units 中型单元变化模式
Scale 1:900 比例 1:900

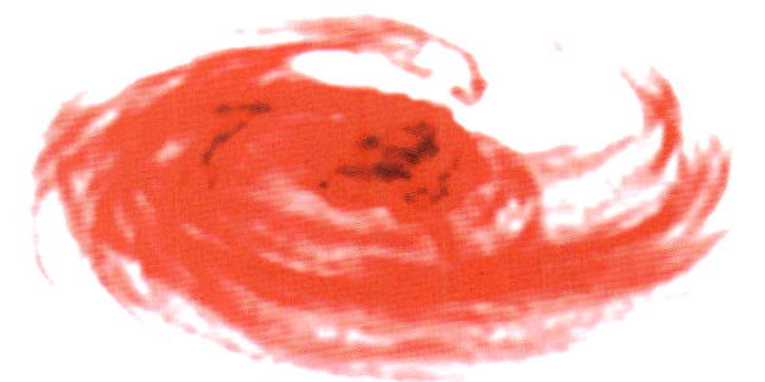

Qi·Parametric Genius Loci
气：参数化场所

tutor: prof. Xu Weiguo
students: Chen Xiaoji,Liang Qiwei,Tao Xiaochen,Xiong Xing,Zhou Shi

我们第一眼看这个球体时感到非常兴奋，因为它包含着如此个人化的不同设计部分，却在整体上显示出一种和谐。我们没有添加新的部分，我们感兴趣的是，用中国哲学中的气来探寻一种解释最终球体的新方法。在开始设计前，我们花了很多时间来仔细研究《易经》（关于中国哲学精华的一本古书）。

《易经》认为，任何事物都由无限小的部分组成，正是这些局部的属性和状态（阴和阳）形成了整体。与建筑相同，它可看作一个在不同因素交互作用下的复杂系统，诸如流线、功能、使用者性别等等。正是这些因素状态在阴阳之间连续相互交换的过程，形成了建筑的动态平衡系统。

根据这本书，气是个抽象概念，可看作一个其组成部分影响下的物体所产生的不可触及的能量场。阴和阳是气的根基（或者驱动力）。这样看来，我们认为建筑的气与西方的“场所”概念是相似的。尤其对于这个巨型球体，因为它是由不同个体但又相互联系的部分组成。

使用由图示球体每个部分的阴阳而得到的参数，我们就能将它的参数化场所可视化。

At first glance of the sphere, we were very excited about the fact that it consists of such individually designed parts and yet demonstrate a certain kind of harmony as a whole. Instead of adding new parts to it, we thought it would be interesting to explore a new method to interpret the final sphere-with the Qi in Chinese philosophy. Before setting out to design, we spent a lot of time studying the Yi Jing (an essential ancient book on Chinese philosophy) carefully.

The Yi Jing believes that everything is consisted of infinite smaller parts, and it is the properties and status (Yin and Yang) of these parts that shapes the whole. Same as in architecture, it can be viewed as a complex system under the interacting forces of different factors, such as circulation, performance, gender of users etc. And it is the continuously interchanging process of the states of these factors between Yin and Yang that have shaped the dynamically-balanced system of architecture.

According to this book, Qi is an abstract concept that can be viewed as an untouchable field of energy produced by an object under the influence of its consisting parts. Yin and Yang are the roots (or driving forces) of Qi. In this sense, we believe that the Qi of an architecture is similar to the western concept of Genius Loci. It is especially the case with the mega sphere, as it is consisted of different individual yet interrelated parts.

And with the parameters we obtain by mapping the Yin and Yang of each part of the sphere, we can visualize the parametric Genius Loci of it.

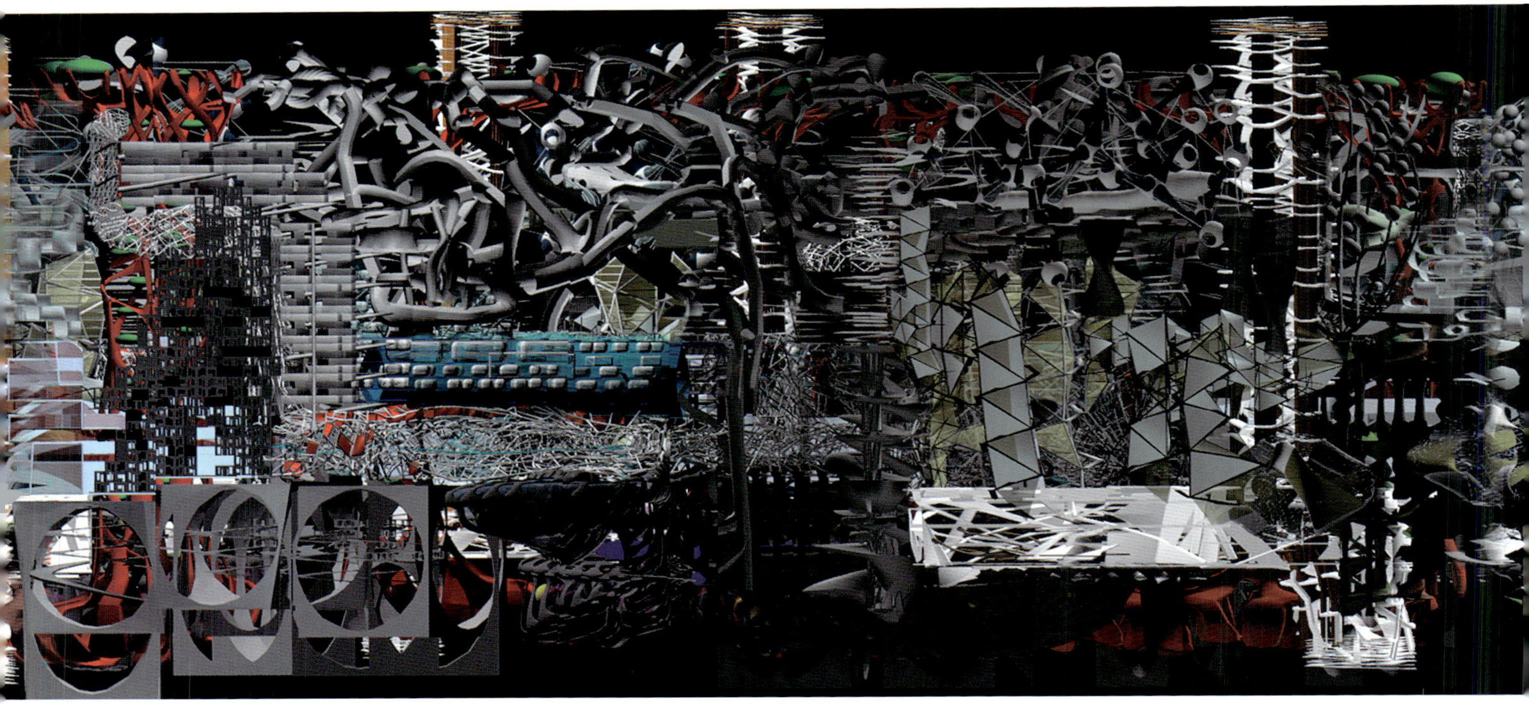

unwrap the surface of the sphere

Yang 陽 —

Yin 陰 – –

Upper-gram
（上卦）
development tendency

Lower-gram
（下卦）
original property

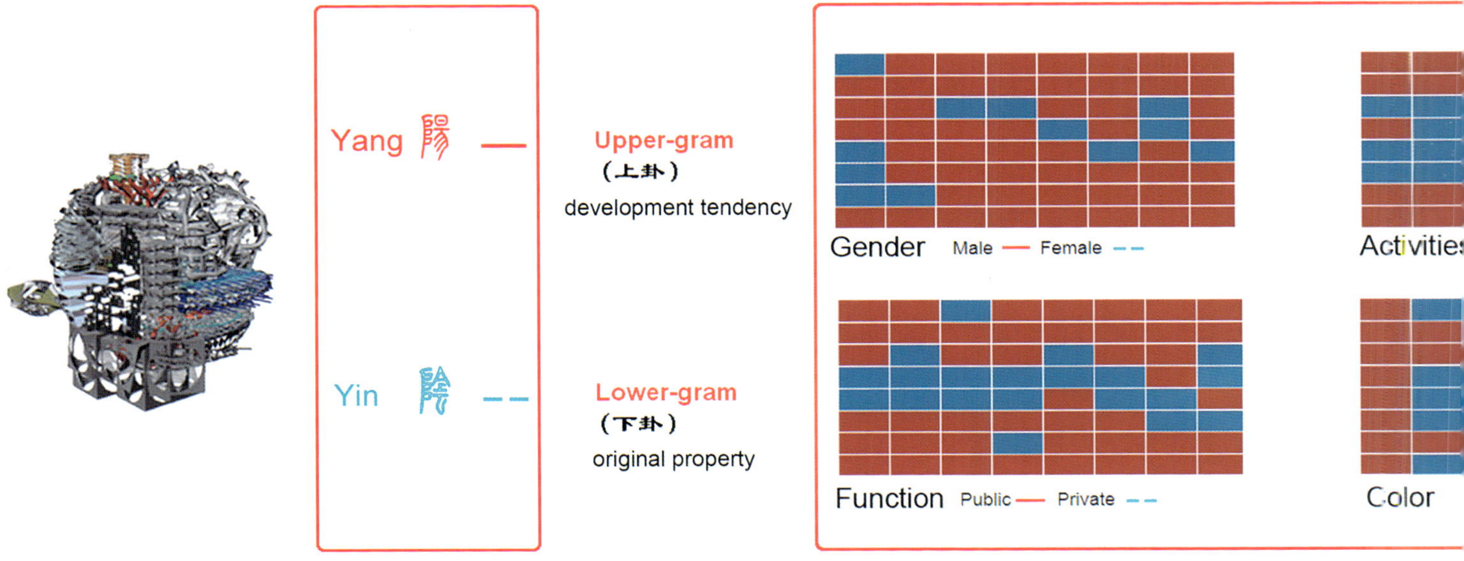

estimate the Yin or Yang of the six lines: Yaos(爻) of every zone

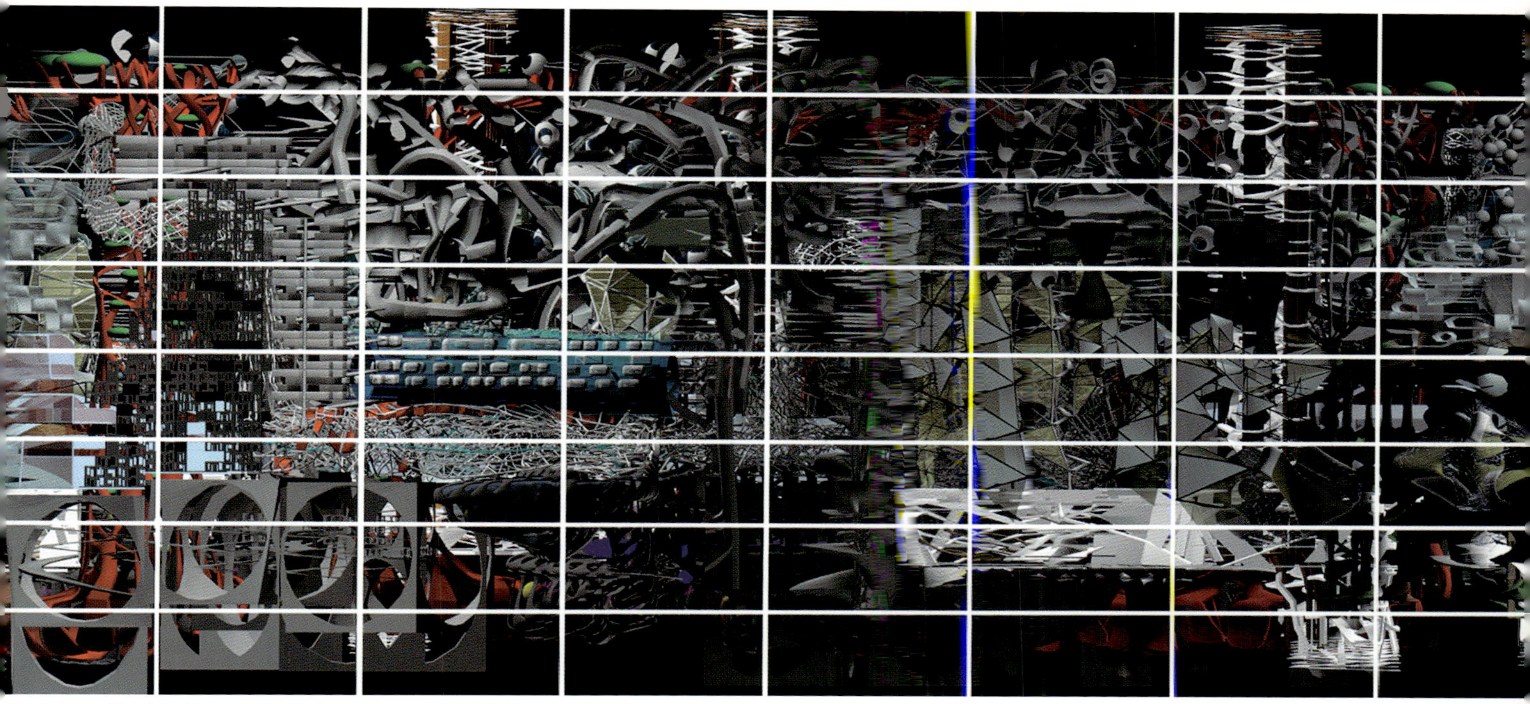

divide the surface into 64 zones

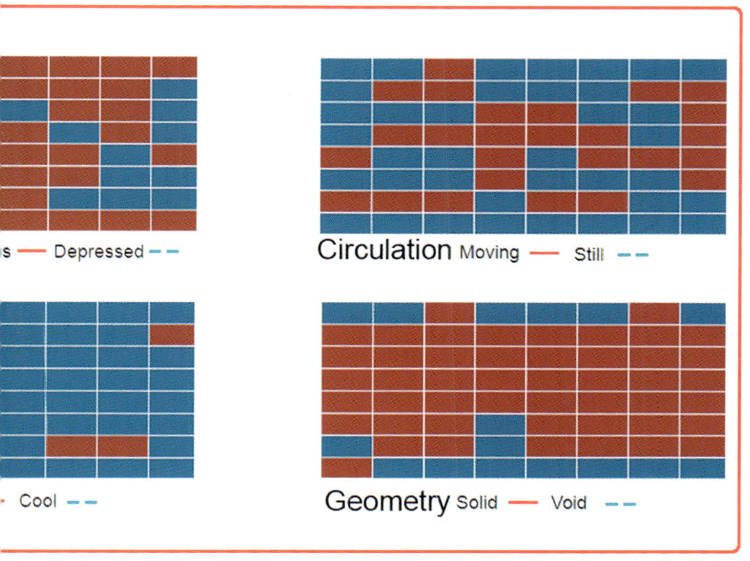

array the six Yaos together, we get the total Gua(卦)of every zone of the sphere

transforming logic

The Yin-Yang parameters also affect the way they array to form Qi according to the majority of the Yin or Yang in 64 zones.

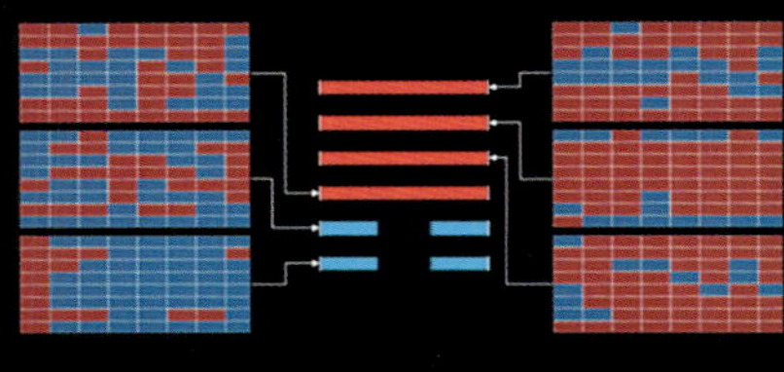

```
            siny=arrNormal(0)(0)-arrNormal(1)(0)
            cosy=sqr(1-siny*siny)
            sinx=(arrNormal(1)(1)-arrNormal(0)(1))/cosy
            cosx=(arrNormal(1)(2)-arrNormal(0)(2))/cosy

            arrMatrix(0,0) = cosy
            arrMatrix(0,1) = 0
            arrMatrix(0,2) = -siny
            arrMatrix(0,3) = arrNormal(0)(0)

            arrMatrix(1,0) = sinx*siny
            arrMatrix(1,1) = cosx
            arrMatrix(1,2) = sinx*cosy
            arrMatrix(1,3) = arrNormal(0)(1)

            arrMatrix(2,0) = cosx*siny
            arrMatrix(2,1) = -sinx
            arrMatrix(2,2) = cosx*cosy
            arrMatrix(2,3) = arrNormal(0)(2)

            arrMatrix(3,0) = 0
            arrMatrix(3,1) = 0
            arrMatrix(3,2) = 0
            arrMatrix(3,3) = 1
            strU=copyUnit (strUnit,i,j)
            Rhino.TransformObject strU, arrMatrix
        Next
    Next
End Sub

Randomize
'Dim arrProperties(5,7)
    arrProperties(0,0)=Array(1,0,0,0,0,0,0,0)
    arrProperties(0,1)=Array(1,0,0,0,0,0,0,0)
    arrProperties(0,2)=Array(1,0,0,0,0,0,0,0)
    arrProperties(0,3)=Array(1,0,0,0,0,0,0,0)
    arrProperties(0,4)=Array(1,0,0,0,0,0,0,0)
    arrProperties(0,5)=Array(1,0,0,0,0,0,0,0)
    arrProperties(0,6)=Array(1,0,0,0,0,0,0,0)
    arrProperties(0,7)=Array(1,0,0,0,0,0,0,0)
```

Yang----expand the shape

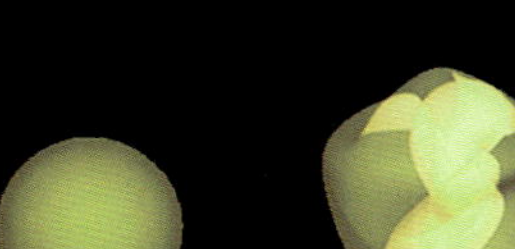

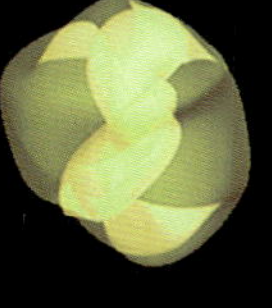

Yin-----contract the shape

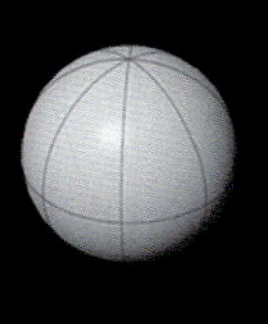

Prototype of a static status of the Qi of the sphere

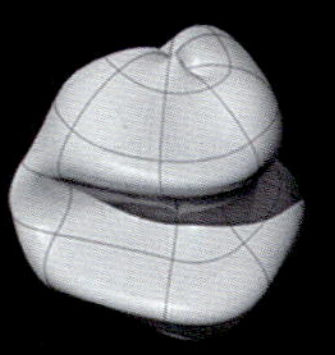

Transform phase 1

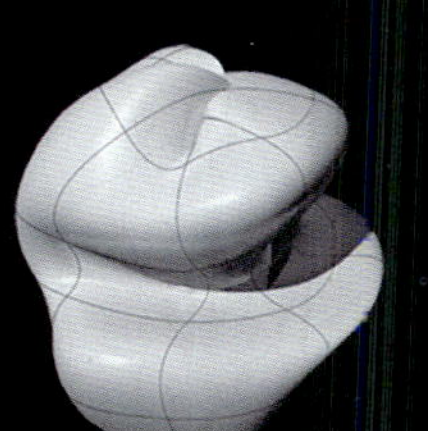

Transform phase 2

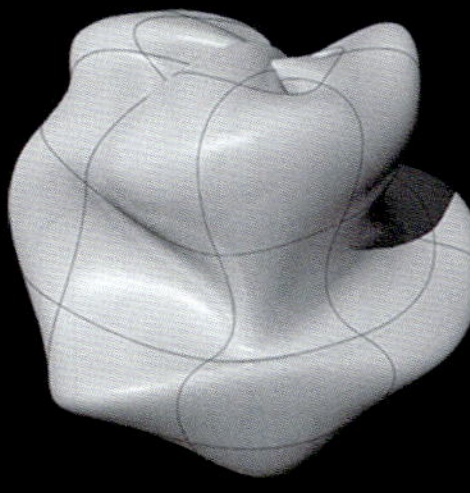

Transform phase 3

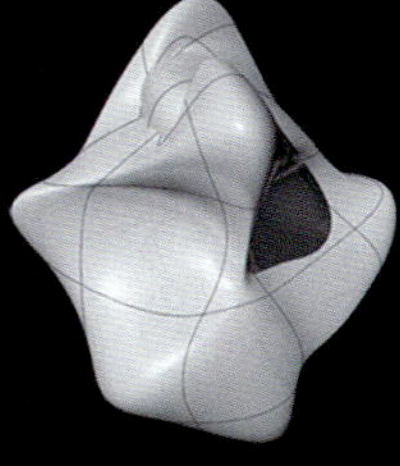

Transform phase 4

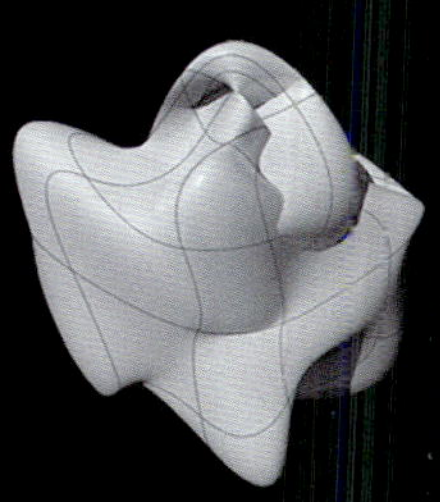

Transform phase 5

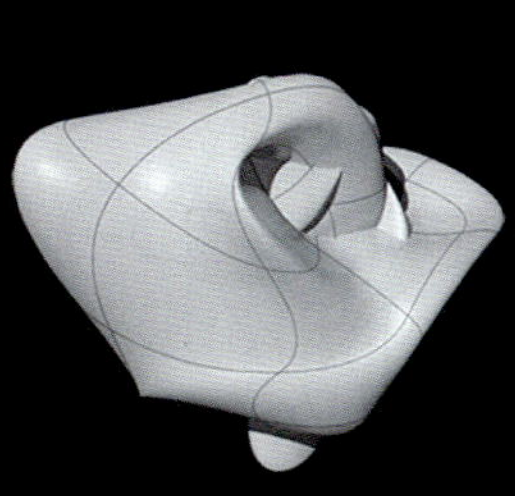

Transform phase 6

Arrayed form with other units